# Alternatives FOR LANDMINE DETECTION

acqueline MacDonald
J.R. Lockwood
John McFee
Thomas Altshuler
Thomas Broach
Lawrence Carin
Russell Harmon
Carey Rappaport
Waymond Scott
Richard Weaver

RAND
Science and Technology Policy Institute
Prepared for the Office of Science and Technology Policy

The research described in this report was conducted by RAND's Science and Technology Policy Institute for the Office of Science and Technology Policy.

**Library of Congress Cataloging-in-Publication Data**

MacDonald, Jacqueline/
Alternatives for landmine detection / Jacqueline MacDonald, J.R. Lockwood.
p. cm.
Includes bibliographical references.
"MR-1608."
ISBN 0-8330-3301-8 (pbk.)
1. Land mines. 2. Explosives—Detection. I. Lockwood, J. R. II. Rand Corporation. III.Title.

UG490 .M337 2003
623' .26—dc21

2002155750



*Cover design by Tanya Maiboroda*



Published 2003 by RAND
1700 Main Street, P.O. Box 2138, Santa Monica, CA 90407-2138
1200 South Hayes Street, Arlington, VA 22202-5050
201 North Craig Street, Suite 202, Pittsburgh, PA 15213-1516
RAND URL: http://www.rand.org/
To order RAND documents or to obtain additional information, contact Distribution Services: Telephone: (310) 451-7002;
Fax: (310) 451-6915; Email: order@rand.org

# PREFACE

This report assesses the potential for innovative mine detection technologies to speed clearance of the 45–50 million landmines around the world. The Office of Science and Technology Policy commissioned the report because of concerns about the slow pace of humanitarian demining.

Numerous studies have compared and evaluated the performance of mine detection technologies—those currently fielded and those under development. This report is unique because it focuses entirely on close-in detection of antipersonnel mines and leverages expertise of two groups of prominent experts including (1) specialists on the cutting edge of the latest technologies and (2) researchers and program managers with long-standing and broad experience in mine detection. Those in the first group wrote background papers describing the most recent research on each innovative detection technology; these papers are included in the appendixes of this report. Those in the second group (who are listed as coauthors of this report) reviewed the background papers and assessed the relative potential of the different technologies. The main report synthesizes the conclusions of this expert group as well as the results of the Science and Technology Policy Institute's (S&TPI's) review of mine detection literature.

The main report of this book was written to be accessible to a wide audience, including federal policymakers, the science community, nongovernmental organizations involved in humanitarian demining, and the general public. The appendixes provide additional technical details on specific mine detection technologies and will be of interest primarily to the science community. RAND is grateful to the authors and the corporate, academic, and governmental groups that allowed us to reproduce this important work in our study.

## THE SCIENCE AND TECHNOLOGY POLICY INSTITUTE

Originally created by Congress in 1991 as the Critical Technologies Institute and renamed in 1998, the Science and Technology Policy Institute is a federally funded research and development center sponsored by the National Science Foundation and managed by RAND. The institute's mission is to help improve public policy by conducting objective, independent research and analysis on policy issues that involve science and technology. To this end, the institute

- supports the Office of Science and Technology Policy and other Executive Branch agencies, offices, and councils;
- helps science and technology decisionmakers understand the likely consequences of their decisions and choose among alternative policies; and
- helps improve understanding in both the public and private sectors of the ways in which science and technology can better serve national objectives.

In carrying out its mission, the institute consults broadly with representatives from private industry, institutions of higher education, and other nonprofit institutions.

Inquiries regarding S&TPI may be directed to the addresses below.

Helga Rippen
Director, S&TPI

**Science and Technology Policy Institute**

RAND
1200 S. Hayes St.
Arlington, VA 22202-5050

Phone: (703) 413-1100, x5574
Web: www.rand.org/scitech/stpi/
Email: stpi@rand.org

# CONTENTS

Preface . . . . . . . . . . . . . . . . . . . . . . . . . . . . . . . . . . iii

Figures . . . . . . . . . . . . . . . . . . . . . . . . . . . . . . . . . . ix

Tables . . . . . . . . . . . . . . . . . . . . . . . . . . . . . . . . . . xiii

Summary . . . . . . . . . . . . . . . . . . . . . . . . . . . . . . . . . xv

Abbreviations . . . . . . . . . . . . . . . . . . . . . . . . . . . . . . xxv

Chapter One
INTRODUCTION . . . . . . . . . . . . . . . . . . . . . . . . . . . . 1
Magnitude of the Antipersonnel Mine Problem . . . . . . . . 2
Design of Antipersonnel Mines . . . . . . . . . . . . . . . . . . . . 3
Limitations of the Conventional Mine Detection Process . . . . . . . . . . . . . . . . . . . . . . . . . 6

Chapter Two
INNOVATIVE MINE DETECTION SYSTEMS . . . . . . . . . . . 15
Method for Evaluating Innovative Mine Detection Systems . . . . . . . . . . . . . . . . . . . . . . . . . 15
Innovative Electromagnetic Detection Systems . . . . . . . . 18
Ground-Penetrating Radar . . . . . . . . . . . . . . . . . . . . . . 18
Electrical Impedance Tomography . . . . . . . . . . . . . . . . 22
X-Ray Backscatter . . . . . . . . . . . . . . . . . . . . . . . . . . . 23
Infrared/Hyperspectral Systems . . . . . . . . . . . . . . . . . . 24
Acoustic/Seismic Systems . . . . . . . . . . . . . . . . . . . . . . . 26
Explosive Vapor Detection Techniques . . . . . . . . . . . . . . 29
Biological Methods . . . . . . . . . . . . . . . . . . . . . . . . . . 30
Chemical Methods . . . . . . . . . . . . . . . . . . . . . . . . . . . 37
Bulk Explosive Detection Techniques . . . . . . . . . . . . . . . 40

Nuclear Quadrupole Resonance . . . . . . . . . . . . . . . . . . 40
Neutron Methods . . . . . . . . . . . . . . . . . . . . . . . . . 42
Innovative Prodders and Probes . . . . . . . . . . . . . . . . . . . 44
Advanced Signal Processing and Signature Modeling . . . . 46

Chapter Three
MULTISENSOR SYSTEM TO IMPROVE MINE DETECTION CAPABILITY . . . . . . . . . . . . . . . . . . . . . . . . . 49
Key Design Considerations . . . . . . . . . . . . . . . . . . . . . 51
Potential for a Multisensor System to Increase Mine Clearance Rate . . . . . . . . . . . . . . . . . . . . . . . . . 54
Current U.S. R&D Investment in Mine Detection Technologies . . . . . . . . . . . . . . . . . . . . . . . . . . . . 56
Recommended Program for Producing an Advanced Multisensor System . . . . . . . . . . . . . . . . . . . . . . . . . 60
Cost of Developing a Multisensor System . . . . . . . . . . . . 61
Summary of Recommendations . . . . . . . . . . . . . . . . . . 64

References . . . . . . . . . . . . . . . . . . . . . . . . . . . . . . . . 65

Author and Task Force Biographies . . . . . . . . . . . . . . . . . . 71

Appendix

A. ELECTROMAGNETIC INDUCTION (PAPER I)
*Yoga Das* . . . . . . . . . . . . . . . . . . . . . . . . . . . . . . . 75

B. ELECTROMAGNETIC INDUCTION (PAPER II)
*Lloyd S. Riggs* . . . . . . . . . . . . . . . . . . . . . . . . . . . 85

C. INFRARED/HYPERSPECTRAL METHODS (PAPER I)
*Brian Baertlein* . . . . . . . . . . . . . . . . . . . . . . . . . . 93

D. INFRARED/HYPERSPECTRAL METHODS (PAPER II)
*John G. Ackenhusen* . . . . . . . . . . . . . . . . . . . . . . . . 111

E. GROUND-PENETRATING RADAR (PAPER I)
*Lawrence Carin* . . . . . . . . . . . . . . . . . . . . . . . . . . 127

F. GROUND-PENETRATING RADAR (PAPER II)
*James Ralston, Anne Andrews, Frank Rotondo, and Michael Tuley* . . . . . . . . . . . . . . . . . . . . . . . . . . 133

G. ACOUSTIC/SEISMIC METHODS (PAPER I)
*James Sabatier* . . . . . . . . . . . . . . . . . . . . . . . . . . . . . . 149

H. ACOUSTIC/SEISMIC METHODS (PAPER II)
*Dimitri M. Donskoy* . . . . . . . . . . . . . . . . . . . . . . . . . . . 155

I. ELECTRICAL IMPEDANCE TOMOGRAPHY
*Philip Church* . . . . . . . . . . . . . . . . . . . . . . . . . . . . . . 161

J. NUCLEAR QUADRUPOLE RESONANCE (PAPER I)
*Andrew D. Hibbs* . . . . . . . . . . . . . . . . . . . . . . . . . . . . 169

K. NUCLEAR QUADRUPOLE RESONANCE (PAPER II)
*Allen N. Garroway* . . . . . . . . . . . . . . . . . . . . . . . . . . . 179

L. X-RAY BACKSCATTER (PAPER I)
*Lee Grodzins* . . . . . . . . . . . . . . . . . . . . . . . . . . . . . . 191

M. X-RAY BACKSCATTER (PAPER II)
*Alan Jacobs and Edward Dugan* . . . . . . . . . . . . . . . . . . . . 205

N. NEUTRON TECHNOLOGIES (PAPER I)
*John E. McFee* . . . . . . . . . . . . . . . . . . . . . . . . . . . . . 225

O. NEUTRON TECHNOLOGIES (PAPER II)
*David A. Sparrow* . . . . . . . . . . . . . . . . . . . . . . . . . . . . 239

P. ELECTROCHEMICAL METHODS (PAPER I)
*Timothy M. Swager* . . . . . . . . . . . . . . . . . . . . . . . . . . . 245

Q. ELECTROCHEMICAL METHODS (PAPER II)
*Thomas F. Jenkins, Alan D. Hewitt, and*
*Thomas A. Ranney* . . . . . . . . . . . . . . . . . . . . . . . . . . . . 255

R. BIOLOGICAL SYSTEMS (PAPER I)
*Robert S. Burlage* . . . . . . . . . . . . . . . . . . . . . . . . . . . . 265

S. BIOLOGICAL SYSTEMS (PAPER II)
*Jerry J. Bromenshenk, Colin B. Henderson, and*
*Garon C. Smith* . . . . . . . . . . . . . . . . . . . . . . . . . . . . . 273

T. CANINE-ASSISTED DETECTION
*Håvard Bach and James Phelan* . . . . . . . . . . . . . . . . . . . . 285

U. SIGNAL-PROCESSING AND SENSOR FUSION METHODS
(PAPER I)
*Leslie Collins* . . . . . . . . . . . . . . . . . . . . . . . . . . . . . . 301

V. SIGNAL-PROCESSING AND SENSOR FUSION METHODS (PAPER II)
*Paul Gader* . . . . . . . . . . . . . . . . . . . . . . . . . . . . . . . . . . . 311

W. CONTACT METHODS
*Kevin Russell* . . . . . . . . . . . . . . . . . . . . . . . . . . . . . . . . . . 327

# FIGURES


1.1. Mine Victims at the Red Cross Limb-Fitting Center in Kabul . . . . . . . . . . . . . . . . . . . . . . . . . . . 3
1.2. Blast Mine . . . . . . . . . . . . . . . . . . . . . . . . . . . . . . . . . 4
1.3. Fragmentation Mine . . . . . . . . . . . . . . . . . . . . . . . . . 5
1.4. At Work in a Mine Detection Lane in Kosovo . . . . . . . 6
1.5. Example (Hypothetical) ROC Curves . . . . . . . . . . . . . 8
1.6. Performance of 29 Commercially Available EMI Mine Detectors in Clay Soil in Cambodia . . . . . . . . . . . . . 11
1.7. Performance of 29 Commercially Available EMI Mine Detectors in Iron-Rich Laterite Soil in Cambodia . . . . 12
2.1. Images of Landmines Produced by GPR System . . . . . 19
2.2. Prototype HSTAMIDS System . . . . . . . . . . . . . . . . . . 20
2.3. Infrared Image of Mines . . . . . . . . . . . . . . . . . . . . . . 25
2.4. Amplitude of Surface Vibration of Ground over a Mine and a Blank in Response to Sound Waves . . . . . . . . . . 27
2.5. Range of Concentrations of 2,4-DNT and TNT in Boundary Layer of Air Near Soil Above Landmines . . . 31
2.6. Mine-Detecting Dog in Bosnia . . . . . . . . . . . . . . . . . . 32
2.7. Bacteria Fluorescing in the Presence of TNT . . . . . . . 35
2.8. Nomadics' Sensor for Detecting Explosive Vapors . . . 38
2.9. Prototype Handheld NQR Mine Detector . . . . . . . . . . 41
3.1. Detection Technologies Funded by the U.S. Humanitarian Demining R&D Program in Fiscal Year 2002 . . . . . . . . . . . . . . . . . . . . . . . . . . . 57
B.1. Typical Electromagnetic Induction System . . . . . . . . 86
B.2. ROC Curve Results for Modified AN/PSS-12 Sensor . . . . . . . . . . . . . . . . . . . . . . . . 90

D.1. Successful IR/HS Mine Detection Must Accommodate a Wide Variety of Conditions .................... 112
I.1. EIT Landmine Detector Prototype ............... 162
I.2. Detector Response for an Antitank Mine-Like Object Buried at a Depth of 14 cm ....................... 163
J.1. Present Configuration of the Man-Portable and Vehicle-Mounted NQR Systems ................ 174
K.1. A Quadrupolar Nucleus Slightly Aligned by the Electrostatic Interaction with the Valence Electrons ........................... 182
K.2. Representative NQR Frequencies for Some Explosives and Narcotics ............................. 183
K.3. Chemical Structure of RDX ..................... 184
L.1. High-Resolution Image of a "Butterfly" Antipersonnel Landmine ........................ 192
L.2. AS&E Backscatter Image ....................... 195
L.3. Illustration of the Standard AS&E Backscatter System Adapted to Work in Conjunction with a Water-Jet Hole Drilling System .......................... 196
L.4. Illustration of a Scanning Electron Beam Backscatter System .......................... 196
L.5. Backscatter System Modified to Use Forward Scatter Radiation ........................... 197
L.6. Schematic Diagram of Vaned Collimator Imaging Technique ........................... 198
L.7. Collection Efficiency as a Function of Depth in Inspection Tunnel Using Vaned Collimators ....... 199
L.8. Schematic Diagram of Backscatter Detector and Collimator Configuration to Measure Atomic Number ........................... 200
L.9. Schematic Diagram of Backscatter Detector and Collimator Configuration to Simultaneously Measure Atomic Number and Density ............ 201
L.10. Two Illustrations of X-Ray Imaging ............. 202
L.11. Nominal Program Schedule ..................... 203
M.1. Configuration Used to Acquire Images ............ 209
M.2. Drawing of Field Test LMR Mine Detection System (XMIS) ............................ 209
M.3. LMR Images of TMA-4 Antitank Mine ............ 210
M.4. LMR Images of M-19 Antitank Mine .............. 211

M.5. LMR Images of Surface-Laid TS/50 Antipersonnel Mine . . . . . . . . . . . . . . . . . . . . . . . . 212
M.6. LMR Images of Flush-with-Surface TS/50 Antipersonnel Mine . . . . . . . . . . . . . . . . . . . . . . . . 213
M.7. Image Set No. 1 Obtained During the Fort A. P. Hill Test . . . . . . . . . . . . . . . . . . . . . . . . . . . . . 216
M.8. Image Set No. 2 Obtained During Fort A. P. Hill Test . . . . . . . . . . . . . . . . . . . . . . . . . . . 217
M.9. Image Set No. 3 Obtained During Fort A. P. Hill Test . . . . . . . . . . . . . . . . . . . . . . . . . . . 218
M.10. Components of XMIS . . . . . . . . . . . . . . . . . . . . . . 220
M.11. XMIS in Use on the Fort A. P. Hill Mine Lanes . . . . . . . 220
S.1. ROC Curve for Honeybees Trained to Associate the Smell of DNT with Food . . . . . . . . . . . . . . . . . . . . . 278
V.1. Typical Signal-Processing Algorithm for Landmine Detection . . . . . . . . . . . . . . . . . . . . . . . . . 313
V.2. Results of Signal Processing on DARPA Background Data . . . . . . . . . . . . . . . . . . . . . . . . . . . 315
V.3. Reduction in FAR Obtained from Spatial Processing of GPR Data Collected from a Handheld Unit . . . . . . . . . 316
V.4. Classification Results on the Acoustics Data . . . . . . . . 317

# TABLES


S.1. Summary of the Detection Technologies Reviewed . . . . . . . . . . . . . . . . . . . . . . . . . . . . . . . xviii
1.1. Time Spent Detecting and Clearing Mines, Unexploded Ordnance, and Scrap in Cambodia, March 1992–October 1998 . . . . . . . . . . . . . . . . . . . . . 10
2.1. Chemical and Physical Methods for Sensing Explosive Vapors . . . . . . . . . . . . . . . . . . . . . . . . . . . . 37
3.1. Sources of False Alarms for Selected Mine Detection Technologies . . . . . . . . . . . . . . . . . . . . . . 50
3.2. Mine Types and Soil Moisture Conditions for Which Selected Mine Detection Technologies Are Best Suited . . . . . . . . . . . . . . . . . . . . . . . . . . . . . . 51
3.3. Estimated Costs and Time Required to Develop a Next-Generation Multisensor Landmine Detector . . . 62
3.4. Itemization of Basic Research Requirements and Costs over the Next Five Years for a Next-Generation Multisensor Mine Detector . . . . . . . . . . . . . . . . . . . 63
D.1. ROC Curve and References for HS/IR Mine Detection Field Tests. . . . . . . . . . . . . . . . . . . . . . . . . . . . . . . 116
D.2. Outline of a Five-to-Seven-Year, $31 Million Program Leading to Robust, Satisfactory IR/HS Minefield Detection . . . . . . . . . . . . . . . . . . . . . . . . . 122
F.1. Data Collection Matrix . . . . . . . . . . . . . . . . . . . . . . . 138
F.2. Other Recommendations . . . . . . . . . . . . . . . . . . . . . 140
I.1. Suggested R&D Program . . . . . . . . . . . . . . . . . . . . . . 167
J.1. Present Signal-to-Noise Ratio for a 1-Second NQR Measurement of 100 g of Explosive . . . . . . . . . . . . . 171

J.2. Example PD and PFA Operating Points 2 to 7 for NQR SNR . . . . . . . . . . 171
J.3. Projected Results for Detection of Low-Metal Landmines at Aberdeen and Yuma Proving Grounds . . . . . . . . . . 173
J.4. Comparison of Two Possible Combinations of NQR with HSTAMIDS for Low-Metal Mines Measured in Off-Road Conditions . . . . . . . . . . 177

# SUMMARY

Antipersonnel mines remain a significant international threat to civilians despite recent intense efforts by the United States, other developed countries, and humanitarian aid organizations to clear them from postconflict regions. Mines claim an estimated 15,000–20,000 victims per year in some 90 countries. They jeopardize the resumption of normal activities—from subsistence farming to commercial enterprise—long after periods of conflict have ceased. For example, in Afghanistan during 2000, mines claimed 150–300 victims per month, half of them children. Although most of these mines were emplaced during the Soviet occupation of Afghanistan (from 1979 to 1988), they continue to pose a serious risk to returning refugees and have placed vast tracts of farmland off limits. The United States currently invests about $100 million annually in humanitarian mine clearance—the largest commitment of any country. Despite this investment and the funding from many other developed nations and nongovernmental organizations, at the current rate clearing all existing mines could take 450–500 years.

This report addresses the following questions:

- What innovative research and development (R&D) is being conducted to improve antipersonnel mine detection capabilities?
- What is the potential for each innovative technology to improve the speed and safety of humanitarian demining?
- What are the barriers to completing development of innovative technologies?

- What funding would be required, and what are the options for federal investments to foster development of promising mine detection technologies?

We focus on close-in detection of antipersonnel mines rather than on airborne or other remote systems for identifying minefields.

The report was written by RAND S&TPI staff and a task force of eight experts in mine detection from universities and U.S. and Canadian government agencies. In addition, 23 scientists provided background papers with details on specific mine detection technologies; these papers are published in this report as separate appendixes.

## LIMITATIONS OF CONVENTIONAL MINE DETECTION TECHNOLOGIES

The tools available to mine detection teams today largely resemble those used during World War II. A deminer is equipped with a handheld metal detector and a prodding device, such as a pointed stick or screwdriver. The demining crew first clears a mined area of vegetation and then divides it into lanes of about a meter wide. A deminer then slowly advances down each lane while swinging the metal detector low to the ground. When the detector signals the presence of an anomaly, a second deminer probes the suspected area to determine whether it contains a buried mine.

The overwhelming limitation of the conventional process is that the metal detector finds every piece of metal scrap, without providing information about whether the item is indeed a mine. For example, of approximately 200 million items excavated during humanitarian demining in Cambodia between 1992 and 1998, only about 500,000 items (less than 0.3 percent) were antipersonnel mines or other explosive devices. The large number of false alarms makes humanitarian mine detection a slow, dangerous, and expensive process. Every buried item signaled by the detector must be investigated manually. Prodding with too much force, or failure to confirm the presence of a mine during probing, can lead to serious injury or death. Adjusting a conventional detector to reduce the false alarm rate results in a simultaneous decrease in the probability of finding a mine, meaning more mines will be left behind when the demining operation is completed. For humanitarian demining, trading off reductions in false alarms for reductions in the likelihood of finding buried mines is unacceptable.

## CAPABILITIES OF INNOVATIVE MINE DETECTION TECHNOLOGIES

Research is under way to develop new detection methods that search for characteristics other than metal content. The aim of these methods is to substantially reduce the false alarm rate while maintaining a high probability of detection, thereby saving time and reducing the chance of injury to the deminer. Table S.1 summarizes these methods. The second column indicates the detection principle on which each is based. The remaining columns summarize the strengths, limitations, and performance potential of each. Chapter Two and the appendixes provide detailed reviews of each technology.

As shown in Table S.1, no single mine detection technology can operate effectively against all mine types in all settings. For example, nuclear quadrupole resonance can find mines containing the explosive cyclotrimethylenenitramine (known as royal demolition explosive [RDX]) relatively quickly, but it is slow in confirming the presence of trinitrotoluene (TNT). Acoustic mine detection systems have demonstrated very low false alarm rates, but they cannot find mines buried at depths greater than about one mine diameter. Chemical vapor sensors can find plastic mines in moist soils, but they have difficulty locating metal mines in dry environments.

Given the limitations of individual sensor technologies, major breakthroughs in mine detection capability are likely to occur only with the development of a multisensor system. The multisensor system we envision would combine two or more of the technologies listed as "promising" in Table S.1 and would leverage advanced algorithms that would process the raw signals in concert to determine whether they are consistent with known mine characteristics. Rather than bringing together two commercially available technologies to form the combined sensor platform, the technology optimization and integration would occur at the design stage, and the development of

**Table S.1**

**Summary of the Detection Technologies Reviewed**

| Technology | Operating Principle | Strengths | Limitations | Potential for Humanitarian Mine Detection |
|---|---|---|---|---|
| **Electromagnetic** | | | | |
| Electromagnetic induction | Induces electric currents in metal components of mine | Performs in a range of environments | Metal clutter; low-metal mines | Established technology |
| Ground-penetrating radar | Reflects radio waves off mine/soil interface | Detects all anomalies, even if nonmetal | Roots, rocks, water pockets, other natural clutter; extremely moist or dry environments | Established technology |
| Electrical impedance tomography | Determines electrical conductivity distribution | Detects all anomalies, even if nonmetal | Dry environments; can detonate mine | Unlikely to yield major gains |
| X-ray backscatter | Images buried objects with x rays | Advanced imaging ability | Slow; emits radiation | Unlikely to yield major gains |
| Infrared/hyperspectral | Assesses temperature, light reflectance differences | Operates from safe standoff distances and scans wide areas quickly | Cannot locate individual mines | Not suitable for close-in detection |
| **Acoustic/Seismic** | Reflects sound or seismic waves off mines | Low false alarm rate; not reliant on electromagnetic properties | Deep mines; vegetation cover; frozen ground | Promising |

**Table S.1—continued**

| Technology | Operating Principle | Strengths | Limitations | Potential for Humanitarian Mine Detection |
|---|---|---|---|---|
| **Explosive Vapor** | | | | |
| Biological (dogs, bees, bacteria) | Living organisms detect explosive vapors | Confirms presence of explosives | Dry environments | Basic research needed to determine potential (though dogs are widely used) |
| Fluorescent | Measures changes in polymer fluorescence in presence of explosive vapors | Confirms presence of explosives | Dry environments | Basic research needed to determine operational potential |
| Electrochemical | Measures changes in polymer electrical resistance upon exposure to explosive vapors | Confirms presence of explosives | Dry environments | Basic research needed to determine whether detection limit can be reduced |
| Piezoelectric | Measures shift in resonant frequency of various materials upon exposure to explosive vapors | Confirms presence of explosives | Dry environments | Basic research needed to determine whether detection limit can be reduced |
| Spectroscopic | Analyzes spectral response of sample | Confirms presence of explosives | Dry environments | Basic research needed to determine whether detection limit can be reduced |

**Table S.1—continued**

| Technology | Operating Principle | Strengths | Limitations | Potential for Humanitarian Mine Detection |
|---|---|---|---|---|
| **Bulk Explosives** | | | | |
| Nuclear quadrupole resonance | Induces radio frequency pulse that causes the chemical bonds in explosives to resonate | Identifies bulk explosives | TNT; liquid explosives; radio frequency interference; quartz-bearing and magnetic soils | Promising |
| Neutron | Induces radiation emissions from the atomic nuclei in explosives | Identifies the elemental content of bulk explosives | Not specific to explosives molecule; moist soil; ground-surface fluctuations | Unlikely to yield major gains |
| **Advanced Prodders/ Probes** | Provide feedback about nature of probed object and amount of force applied by probe | Could deploy almost any type of detection method | Hard ground, roots, rocks; requires physical contact with mine | Promising |

algorithms for advanced signal processing would be an integral part of the process. The result would be a single, highly sensitive, and performance-optimized detection system that provides one specific signal to the operator. The Army countermine program currently is developing a dual-sensor system that combines separate electromagnetic induction (EMI) and ground-penetrating radar (GPR) technologies as part of a single operational platform known as the Handheld Standoff Mine Detection System (HSTAMIDS). However, HSTAMIDS does not use advanced signal processing. Rather, the operator receives two separate outputs: one from the EMI device and one from the GPR. This dual-sensor system does not make optimal use of the totality of information available from the combined sensors.

Advances in signal processing and understanding of single-sensor systems make the development of a multisensor system with a single signal possible in principle. Preliminary research has shown the potential for multisensor systems to reduce the number of false alarms by as much as a factor of 12. However, additional research is needed to establish a comprehensive technical basis for the design of such a system. Based on the time and costs required to create HSTAMIDS ($73 million over 15 years), we estimate that the new multisensor system would require a total investment of $135 million. Currently, the United States is not funding the necessary research. In 2002, the United States invested $2.7 million for close-in mine detection R&D for humanitarian demining. Of this amount, nearly $2.0 million went to making incremental improvements to existing EMI and GPR systems, and the rest funded research on explosive chemical vapor detection systems. No funding was allocated toward research that would lead to the development of an integrated multisensor system for humanitarian demining.

At the outset of this project, the Office of Science and Technology Policy asked RAND S&TPI whether development of an innovative mine detection system could enable mine clearance to advance 10 times faster than is currently possible. A multisensor system could reduce the false alarm rate by a factor of 10 or more. However, gains in mine clearance speed are not directly proportional to reductions in the false alarm rate because a substantial portion of the total clearance time is spent on site preparation activities, such as vegetation clearance. Very limited research has been conducted to date to analyze actual mine clearance data for determining what gains are theoretically possible with improved detection systems. The existing, limited research predicts that a system that eliminated 99 percent of false alarms would improve overall clearance rates by 60–300 percent of current rates, depending on the amount of vegetation present. Such gains would save billions to tens of billions of dollars in the total cost expected to clear all mines and would spare a large number of deminer and civilian lives. Pursuing development of an advanced multisensor system is worthwhile, even if order-of-magnitude decreases in clearance time are not possible with improved detection technology alone.

## RECOMMENDATION: INITIATE AN R&D PROGRAM TO DEVELOP A MULTISENSOR SYSTEM

We recommend that the federal government undertake an R&D effort to develop a multisensor mine detection system. The first step in developing the program should be a short, preliminary study (costing less than $1 million) to consolidate existing theoretical and empirical research related to multisensor systems and signal processing. This preliminary study would be used to develop a blueprint for the R&D needed to produce a prototype system. We estimate that initial prototype development would cost approximately $60 million. The program should address the following four broad areas:

- algorithmic fusion of data from individual sensors (to develop the theory necessary to support an advanced multisensor system), funded at approximately $2.0–3.2 million per year;
- integration of component technologies (to address system engineering issues associated with combining multiple sensors as part of a single-sensor platform), funded at approximately $1.25–2.00 million per year;
- methods for detecting the chemical components of explosives (to further develop components of the multisensor system that would search for explosives rather than for the mine casing and mechanical components), supported at approximately $2.5–4.0 million per year; and

- techniques for modeling how soil conditions in the shallow subsurface environment affect various mine sensors (to allow predictions of integrated sensor system performance across the broad range of natural environments in which mines occur), funded at $500,000–800,000 per year.

Depending on the amount of resources invested in this research, a prototype multisensor system could be available within seven years. Once the prototype is developed, additional allocations totaling approximately $135 million will be needed to fund the engineering and development of an optimal, deployable system.

The benefits of a program to develop an advanced, multisensor system would include more rapid capability to help restore stability to postconflict regions, such as Afghanistan; more mines cleared per U.S. dollar spent on humanitarian demining; fewer deminer and civilian casualties; and utility to military countermine operations. In addition, the results of R&D on advanced signal processing and sensor fusion would be transferable to other applications in environmental, geophysical, medical, and other sciences.

# ABBREVIATIONS

| | |
|---|---|
| 2ADNT | 2-amino-4,6-dinitrotuluene |
| 4ADNT | 4-amino-2,6-dinitrotuluene |
| A/S | Acoustic-to-seismic |
| AFRL | Air Force Research Laboratory |
| AP | Antipersonnel |
| APG | Aberdeen Proving Ground |
| ARO | Army Research Office |
| AS&E | American Science and Engineering Inc. |
| ASTAMIDS | Airborne Standoff Minefield Detection System |
| AT | Antitank |
| BTI | Bubble Technology Industries |
| CBI | Compton backscatter imaging |
| CCMAT | Canadian Centre for Mine Action Technologies |
| CECOM | Communications-Electronics Command |
| CMAC | Cambodian Mine Action Centre |
| COBRA | Coastal Battlefield Reconnaissance and Analysis |
| CompB | Composition B |
| COTS | Commercial-off-the-shelf |

| | |
|---|---|
| CSS | Coastal Systems Station |
| CW | Continuous wave |
| DARPA | Defense Advanced Research Projects Agency |
| DC | Direct current |
| DERA | Defence Evaluation and Research Agency |
| DNT | Dinitrotoluene |
| DoD | U.S. Department of Defense |
| DRDC | Defence R&D Canada |
| DRES | Defence Research Establishment–Suffield |
| EIT | Electrical impedance tomography |
| EM | Electromagnetic |
| EMI | Electromagnetic induction |
| EO | Electro-optical |
| ERDC-CRREL | Engineer Research and Development Center, Cold Regions Research and Engineering Laboratory (U.S. Army) |
| FA | False alarm |
| FAA | Federal Aviation Administration |
| FAR | False alarm rate |
| FNA | Fast neutron analysis |
| G | Gauss |
| GHz | Gigahertz |
| GICHD | Geneva International Centre for Humanitarian Demining |
| GPR | Ground-penetrating radar |
| GPSAR | Ground-penetrating synthetic aperture radar |
| GSTAMIDS | Ground Standoff Mine Detection System |
| HD | Humanitarian demining |

| | |
|---|---|
| HMM | Hidden Markov model |
| HMX | Cyclotetramethylenetetranitramine |
| HPM | High-power microwave |
| HS | Hyperspectral |
| HSTAMIDS | Handheld Standoff Mine Detection System |
| ICBL | International Campaign to Ban Landmines |
| IDA | Institute for Defense Analyses |
| ILDP | Improved Landmine Detector Program |
| IPPTC | International Pilot Project for Technology Cooperation |
| IR | Infrared |
| JUXOCO | Joint Unexploded Ordnance Coordination Office |
| keV | Kilo-electron volt |
| kHz | Kilohertz |
| kVp | Kilovolt peak |
| kG | Kilogauss |
| LAMD | Lightweight Airborne Minefield Detection |
| LDV | Laser Doppler vibrometer |
| LED | Light-emitting diode |
| LM | Low metallic |
| LMR | Lateral migration radiography |
| LWIR | Long-wave infrared |
| μm | Micrometer |
| mA | Milliamp |
| MDD | Mine dog detection |
| MeV | Mega-electron volt |

| | |
|---|---|
| MHz | Megahertz |
| MLP | Multilayer perception |
| MMDS | Microbial Mine Detection System |
| MRI | Magnetic resonance imaging |
| mS | MilliSiemens |
| MURI | Multidisciplinary University Research Initiative |
| MWIR | Mid-wave infrared |
| NASA | National Aeronautics and Space Administration |
| NATO | North Atlantic Treaty Organization |
| NCPA | National Center for Physical Acoustics |
| NIR | Near infrared |
| NMR | Nuclear magnetic resonance |
| NQR | Nuclear quadrupole resonance |
| NRL | Naval Research Laboratory |
| NVESD | Night Vision and Electronic Sensors Directorate (U.S. Army) |
| ORNL | Oak Ridge National Laboratories |
| OSTP | Office of Science and Technology Policy |
| PCMCIA | Personal Computer Memory Card International Association |
| PD | Probability of detection |
| PFA | Probability of false alarm |
| PIN | Positive-intrinsic-negative |
| PM | Photo-multiplier |
| ppb | Parts per billion |
| ppq | Parts per quadrillion |
| ppm | Parts per million |

| | |
|---|---|
| ppt | Parts per trillion |
| PY | Person year |
| QM | Quantum Magnetics |
| R&D | Research and development |
| RDX | Royal demolition explosive [Cyclotrimethylenenitramine] |
| REMIDS | Remote Minefield Detection System |
| REST | Remote Explosive Scent Tracing |
| RF | Radio frequency |
| RFI | Radio frequency interference |
| ROC | Receiver operating characteristic |
| ROM | Rough order of magnitude |
| rpm | Revolutions per minute |
| S&TPI | Science and Technology Policy Institute |
| SAR | Synthetic aperture radar |
| SCR | Signal-to-clutter ratio |
| SERRS | Surface Enhanced Resonance Raman Spectroscopy |
| SIT | Stevens Institute of Technology |
| SNL | Sandia National Laboratories |
| SNR | Signal-to-noise ratio |
| SPIE | International Society for Optical Engineering |
| STAMIDS | Standoff Minefield Detection System |
| SWIR | Short-wave infrared |
| SwRI | Southwest Research Institute |
| T | Tesla |
| TNA | Thermal neutron analysis |

| | |
|---|---|
| TNO | Netherlands Organisation for Applied Scientific Research |
| TNT | Trinitrotoluene |
| TOF | Time of flight |
| UF | University of Florida |
| UHF | Ultra-high frequency |
| UM | University of Montana |
| UN | United Nations |
| UWB | Ultra-wide band |
| UXO | Unexploded ordnance |
| VMMD | Vehicle-Mounted Mine Detector |
| VNIR | Visible/near infrared |
| WES | Waterways Experimental Station |
| XMIS | X-ray mine imaging system |

Chapter One

# INTRODUCTION

This report assesses innovative technologies for detecting antipersonnel mines at close range. The Office of Science and Technology Policy (OSTP), concerned about the slow pace of mine detection and clearance in postconflict areas, asked the RAND Science and Technology Policy Institute (S&TPI) to assess the performance potential of innovative mine detection systems and to carry out the following tasks, focusing on close-in detection of antipersonnel mines in humanitarian demining operations:

- Identify antipersonnel mine detection technologies currently in the research and development (R&D) stage.
- Evaluate the potential for each technology to improve the speed and safety of humanitarian demining.
- Identify any barriers to completing development of new technologies.
- Provide information on funding requirements to complete development of new methods.
- Recommend options for federal investments to foster development of key technologies.

This report presents the results of RAND S&TPI's review. This chapter provides background information about landmines and the limitations of existing mine detectors. The next chapter evaluates the potential of innovative detection systems to overcome these limitations. Chapter Three recommends directions for a research program to develop an advanced mine detection system.

## MAGNITUDE OF THE ANTIPERSONNEL MINE PROBLEM

OSTP's concern about landmines is motivated by the fact that antipersonnel mines remain a significant threat in many nations despite focused programs by the United Nations and humanitarian organizations to clear them. Landmines claim an estimated 15,000–20,000 victims per year in 90 countries (ICBL, 2001). The U.S. State Department estimates that a total of 45–50 million mines remain to be cleared (U.S. Department of State, 2001). Worldwide, an estimated 100,000 mines are found and destroyed per year (Horowitz et al., 1996). At that rate, clearing all 45–50 million mines will require 450–500 years, assuming no new mines are laid. By some estimates, roughly 1.9 million new mines are emplaced annually, yielding an additional 19 years of mine clearance work every year (Horowitz et al., 1996).[1]

Mines are inexpensive—costing as little as $3 each—but they impose devastating consequences on the affected communities (Andersson et al., 1995). A survey by Andersson et al. (1995) in Afghanistan, Bosnia, Cambodia, and Mozambique found that one in three victims of mine blasts die. Many of the victims are children. For example, in Afghanistan, the survey found that, on average, 17 in 1,000 children had been injured or killed by mines. For those who survived the blasts, the most common injury reported was loss of a leg (see Figure 1.1). Loss of arms, loss of digits, blindness, and shrapnel wounds also occur.

The presence of mines also can cause economic decline (Andersson et al., 1995; Jeffrey, 1996; Cameron et al., 1998). Most victims are males of working age, and often they are unable to return to work. The Andersson et al. survey found that "households with a mine

---

[1]Estimates of the total number of landmines, the number of victims, and the mine clearance rates are highly uncertain because only about 30 countries have conducted formal surveys (ICBL, 2001). The numbers on total remaining mines that we present here are based on recent U.S. State Department surveys. The estimates of the number of victims are from the International Campaign to Ban Landmines, which annually documents landmine injury data. The estimates of number of mines cleared per year are from a report prepared for the Defense Advanced Research Projects Agency (DARPA). References for all these estimates are shown in the text.

RANDMR1608-1.1

SOURCE: Kaboul. Centre orthopédique du CICR. Photographer: MAYER, Till. Copyright © ICRC 01/01/1996.

**Figure 1.1—Mine Victims at the Red Cross Limb-Fitting Center in Kabul**

victim were 40% more likely to report difficulty in providing food for the family." Further, the medical bills for survivors can bankrupt families. Many victims must undergo multiple surgeries. Children who lose limbs require multiple prosthetic devices over their lifetimes. Mines affect not only the victims' families but also the entire community surrounding the mined area. Even the rumor of mine presence can halt all activity in an affected area. For example, in Mozambique, a town of 10,000 was deserted for four years because of a rumor that mines were present; a three-month clearance operation later found only four mines (Vines and Thompson, 1999). The extensive mine contamination of Afghanistan's fertile valleys has reduced agricultural production; Andersson et al. (1995) estimated that without mines, agricultural land use in Afghanistan could increase by 88–200 percent.

## DESIGN OF ANTIPERSONNEL MINES

Antipersonnel mines were first used in World War II to prevent opposing soldiers from clearing antitank mines. The original antipersonnel mines were improvised from hand grenades and simple electric fuses. Since then, mine design has changed substantially. Modern-day mines can deliver blasts of lethal pellets extending in a radius of up to 100 m (Ackenhusen et al., 2001). Some can be scattered by vehicles, helicopters, or low-flying planes (Ackenhusen et al., 2001). Some are designed to resemble toys or other everyday objects, such as pens and watches. At least 350 mine types exist, manufactured by some 50 countries (Vines and Thompson, 1999).

Although hundreds of mine varieties exist, mines generally can be classified as either "blast" or "fragmentation." Blast mines (see Figure 1.2) are buried at shallow depths. They are triggered by pressure, such as from a person stepping on the mine. The weight needed to activate a blast mine typically ranges from 5 to 24 lb (Ackenhusen et al., 2001), meaning the mines are easily triggered by a small child's weight. They cause the affected object (e.g., foot) to blast into fragments, which blast upward and often are the major cause of damage.

Blast mines typically are cylindrical in shape, 2–4 inches in diameter, and 1.5–3.0 inches in height (Horowitz et al., 1996). Generally, they contain 30–200 g of explosives (Ackenhusen et al., 2001). The casing may be made of plastic, wood, or sheet metal. Plastic-encased blast mines are sometimes referred to as "nonmetallic mines," but nearly

RAND*MR1608-1.2*

SOURCE: Ackenhusen et al. (2001).

**Figure 1.2—Blast Mine**

all of them contain some metal parts, usually the firing pin and a spring/washer mechanism, weighing a gram or so (Horowitz et al., 1996).

Fragmentation mines (see Figure 1.3) throw fragments radially outward at high speeds. Most are lethal and can cause multiple casualties at distances of up to 100 m (Ackenhusen et al., 2001). One type of fragmentation mine, known as the "bounding" mine, is buried underground but is propelled upward when activated and explodes a meter above ground, sending lethal fragments in a wide radius. Other types of fragmentation mines are mounted on stakes in the ground or on tree trunks.

All modern fragmentation mines use steel and therefore are readily found by metal detectors (Ackenhusen et al., 2001). However, they often are activated by tripwires; movement at distances of up to 20 m can trigger the mine before it is located by a mine detector. Tripwire clearance therefore is an essential part of demining. Fragmentation mines come in a wide variety of sizes and shapes, from small cylindrical devices containing as little as 3 oz of explosives to larger containers containing as much as 1 lb of explosives.

RAND*MR1608-1.3*

SOURCE: Ackenhusen et al. (2001).

**Figure 1.3—Fragmentation Mine**

## LIMITATIONS OF THE CONVENTIONAL MINE DETECTION PROCESS

The principles of mine detection have changed little since World War II. The typical deminer's tool kit today largely resembles those used more than 50 years ago. It consists of a metal detector, a prodding instrument (such as a stainless steel probe, pointed stick, or screwdriver), and a tripwire "feeler" made of a coat hanger or 14-gauge wire (Carruthers and McFee, 1996). The demining team typically divides a mined area into grids (commonly 100 sq m), splits the grids into lanes (usually 1 m wide), and then slowly advances down the lanes swinging the metal detector close to the ground (see Figure 1.4). When the detector beeps, the deminer probes the suspected area to determine whether the detected object might be a mine. If the object is not a mine, then it is excavated and laid aside. If it is a mine, then it is detonated in place using a variety of methods (shooting it with a gun, for example). Variations on this process occur depending on location. For example, in some cases mine-sniffing dogs augment the metal detectors. In other places, mines are detonated with mechanical flails or rollers in advance of the detection

RAND*MR1608-1.4*

SOURCE: RONCO Consulting Corp., www.roncoconsulting.com.

**Figure 1.4—At Work in a Mine Detection Lane in Kosovo**

crews.[2] But in all cases, handheld metal detectors and probes are critical parts of the operation.

The metal detectors that are the key part of the deminer's tool kit employ the same principles as those first used in World War I and refined during World War II (Ackenhusen et al., 2001). (In fact, the U.S. military has replaced its standard-issue mine detector only once in the past 40 years [GAO, 1996].) The detectors operate via a principle known as electromagnetic induction (EMI). EMI is used for metal detection not only in mine detection but also by everyday hobbyists. Although advancements in electronics have enabled the development of EMI systems that can detect extremely small amounts of metal and that are lighter and easier to operate than their World War II counterparts, significant limitations to this technology remain.

A prototypical EMI detector consists of a single wire coil or, more commonly, a concentric pair of coils attached to a handle. The deminer holds the coil close to the ground and sweeps it slowly around the area being investigated. Electrical current flowing through the first coil, the "transmit coil," induces a time-varying magnetic field in the ground. This primary magnetic field, in turn, induces electrical currents in buried metal objects. The currents from the buried objects create a weaker, secondary magnetic field. The second coil, the "receiver coil," detects changes in voltage induced by the secondary magnetic field. The detector then converts these changes in the electric potential to an audible signal.

The overwhelming limitation of mine detection using EMI is the inability of EMI systems to discriminate mines from nonmine metal clutter. In theory, EMI systems should be able to find nearly all mines because nearly all contain some metal (Horowitz et al., 1996). However, as the detector is adjusted to signal the small mass of metal present in modern plastic blast mines, it also becomes increasingly sensitive to other metal present in the environment (including shrapnel, bottle caps, bullet casings, and other man-made clutter as well as

---

[2]Mechanical mine clearance using rollers and flails alone has proven ineffective even on flat, clear terrain because many mines fail to detonate on first impact from the mechanical device (Blagden, 1996). In addition, mechanical systems often scatter the mines, leaving some intact and highly unstable.

natural metal in rock). The operator therefore must strike a balance between tuning the detector so finely, that it generates an overwhelming number of false positive signals, and not tuning it finely enough, in which case it misses too many mines.

The balance between the two competing objectives of minimizing the number of false alarms and maximizing the number of mines detected is quantified by what is known as a receiver operating characteristic (ROC) curve. A ROC curve plots the probability of finding a buried mine (the probability of detection, or PD) against the probability that a detected item will be a false alarm (the probability of false alarm, or PFA). Both probabilities are plotted as a function of the threshold used to decide whether to make a declaration (e.g., the loudness of the tone produced by an EMI detector), thus defining a curve. Figure 1.5 shows theoretical examples of ROC curves. The ROC curve for a perfect detector would be a right angle (i.e., 100-percent detection at 0-percent false alarm), while that for a detector equivalent to random guessing would be a 45-degree diagonal line.

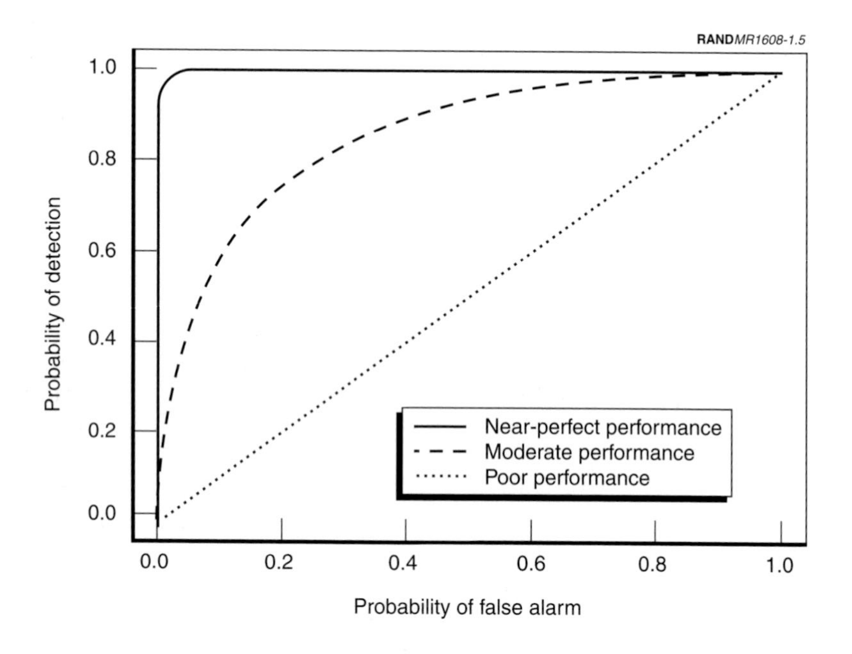


**Figure 1.5—Example (Hypothetical) ROC Curves**

Empirical studies often report detector operating characteristics with a variant of the ROC curve that plots the estimated probability of detection against the false alarm rate (FAR), expressed as the number of false alarms per unit area, rather than against probability of false alarm. This is because, operationally, the false alarm rate is a more natural quantity to consider. It is also more easily calculated because computing the probability of false alarm requires the difficult task of determining how many false alarms could have occurred (Rotondo et al., 1998). Actual ROC curves vary not only with the detector but also with the environment in which the detector is employed, the mine type, and the burial depth.

The high false alarm rate associated with conventional EMI systems accounts for the slow, laborious nature of mine detection (once vegetation and tripwires are cleared). Often, demining teams uncover 100–1,000 inert metal objects for every mine (Hewish and Pengelley, 1997). As an example, Table 1.1 provides data from the Cambodian Mine Action Centre (CMAC) from March 1992 to October 1998. As shown, approximately 90,000 antipersonnel mines were cleared during this time, but 200 million scrap items were excavated in the process. For this period, the probability of false alarm was 0.997 (meaning that for every item detected, there was a 99.7-percent chance that it was scrap and a corresponding 0.3-percent chance that it was a mine or unexploded ordnance [UXO] item). Table 1.1 also shows the time spent digging up scrap items and the time spent digging and neutralizing mines and UXO. In total, 99.6 percent of the time went to excavating scrap items.

Compounding the false alarm problem is that, no matter how careful the EMI operator, EMI systems can still miss mines. A recent international study, known as the International Pilot Project for Technology Cooperation, provided the most comprehensive existing evaluation of all commercially available EMI systems for mine detection (Das et al., 2001). Probabilities of detection varied remarkably by detector, location, and soil type. The best performing detector found 91 percent of the test mines in clay soil (see Figure 1.6), but that same detector found only 71 percent of the mines in laterite (iron-rich) soil (see Figure 1.7). The poorest-performing detector found 11 percent of mines in the clay soil and 5 percent in the laterite soil. By comparison, the UN standard for mine clearance is 99.6 percent. Based on

**Table 1.1**

**Time Spent Detecting and Clearing Mines, Unexploded Ordnance, and Scrap in Cambodia, March 1992–October 1998**

| Type of Item | Total Number of Items Found | Time to Confirm (hours) | Time to Dig (hours) | Time to Neutralize (hours) | Total Time (hours) | Percentage of Total Time |
|---|---|---|---|---|---|---|
| Antitank mines | 961 | 80 | 80 | 80 | 240 | 0.00074 |
| Antipersonnel mines | 89,327 | 7,400 | 7,400 | 7,400 | 22,000 | 0.068 |
| UXO | 452,770 | 38,000 | 38,000 | 38,000 | 110,000 | 0.34 |
| Scrap | 191,737,707 | 16,000,000 | 16,000,000 | 0 | 32,000,000 | 99.6 |

SOURCE: J. McFee, Defence R&D Canada (based on data from CMAC 2000 briefing package, assuming three minutes per item for confirmation, neutralization, or digging).

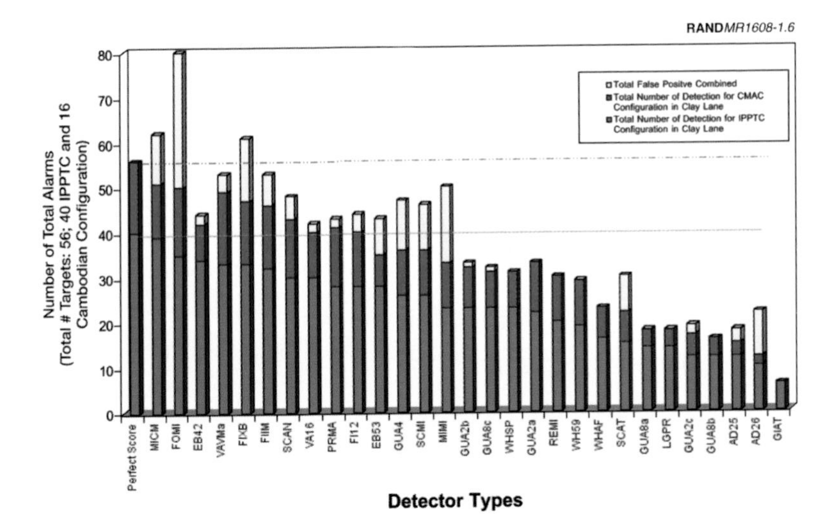


SOURCE: Das et al. (2001).

NOTE: The horizontal axis indexes detector type, while the vertical axis quantifies detector performance against 56 total emplaced objects (subdivided into two configurations of 16 and 40 targets; details on this distinction and on the specific manufacturers and detectors can be found in Das et al., 2001). For each detector, the vertical bar depicts the number of targets detected for each configuration, along with the total number of false alarms. The bar on the far left is for a hypothetical detector with perfect performance.

**Figure 1.6—Performance of 29 Commercially Available EMI Mine Detectors in Clay Soil in Cambodia**

the performance of EMI systems in the International Pilot Project field tests, it appears unlikely that the UN standard is being achieved with current technology, although such quality control measures as surveying areas more than once improve the overall effectiveness.

The high probability of false alarms and imperfect probability of detection of conventional mine detectors make mine detection an occupationally high-risk operation. Data on deminer injury rates are incomplete, but one report indicated that a deminer is killed or maimed for every 1,000–2,000 mines cleared (Blagden, 1996). The amount of time spent investigating false alarms can lead to deminer fatigue and carelessness in investigating identified objects. The imperfect probability of detection means that detection crews may

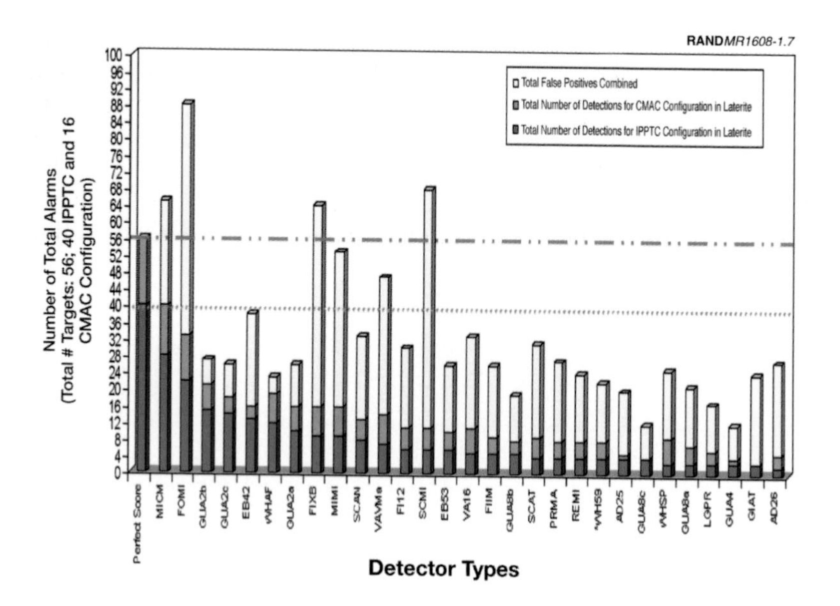


SOURCE: Das et al. (2001).

NOTE: The horizontal axis indexes detector type, while the vertical axis quantifies detector performance against 56 total emplaced objects (subdivided into two configurations of 16 and 40 targets; details on this distinction and on the specific manufacturers and detectors can be found in Das et al., 2001). For each detector, the vertical bar depicts the number of targets detected for each configuration, along with the total number of false alarms. The bar on the far left is for a hypothetical detector with perfect performance.

**Figure 1.7—Performance of 29 Commercially Available EMI Mine Detectors in Iron-Rich Laterite Soil in Cambodia**

be exposed to missed mines. A survey of demining accidents in nine countries found that injuries and fatalities occur most commonly during the excavation stage. Such casualties account for 35 percent of all the demining accidents (Office of the Assistant Secretary of Defense for Special Operations/Low Intensity Conflict, 2000). Presumably, if the deminer thought there was a high likelihood that a detected object was a mine, he or she would be more careful during excavation. The second leading cause of injury, accounting for 24 percent of accidents, was missed mines.

Further improvements in EMI performance are possible with the development of computer algorithms that can help the operator use the return signals to improve discrimination of mines from clutter. However, as is evident from the length of time that EMI has been in use, this technology is very mature, and major breakthroughs are highly unlikely. The next chapter describes research to develop alternative mine detection systems that could be used either alone or in conjunction with EMI systems. The goal in developing alternatives is to decrease the probability of false alarm and increase the probability of detection.

Chapter Two

# INNOVATIVE MINE DETECTION SYSTEMS

Researchers in physical, chemical, and biological sciences are studying and developing methods that could reduce the false alarm rate and maintain or increase the probability of detection for mine clearance. New detection concepts involve searching for characteristics other than mine metal content. A variety of techniques that exploit properties of the electromagnetic spectrum are being explored. In addition, research is under way to develop methods based on acoustics of the mine casing. Biological and chemical methods for detecting explosive vapors also are being explored, as are methods for detecting bulk explosives based on chemical properties. Work also is under way to develop advanced prodders that provide information about the physical characteristics of the object being investigated.

This chapter describes the difficulties of predicting the performance of innovative mine detection methods and our method for assessing the potential of innovative systems. It then describes each type of innovative technology and evaluates its potential to improve on existing EMI detection systems. Table S.1 in the Summary provides an overview of the technology evaluations.

## METHOD FOR EVALUATING INNOVATIVE MINE DETECTION SYSTEMS

Predicting the potential for an innovative mine detection system to reduce the false alarm rate and increase the probability of detection is an inherently difficult task. Research is being conducted by a myriad of universities, government institutions, and private companies, with different projects in various stages of development. However, most of the technologies have not yet been field tested. This makes it virtually impossible to assess operating characteristics (for example, ROC curves) with any specificity, thus precluding defensible quantitative performance comparisons.

Compounding the general lack of data to support quantitative evaluations is a lack of comparability of the data that are available. The performance of a landmine detection system depends on the types and depths of mines present, the environment in which the system is operated, and the human operator. For detectors that locate buried objects, such mine properties as size, shape, and metallic content substantially affect detector performance, as do the placements (depth and orientation) of the mines. Detectors that search for explosives can be sensitive to the type of explosive contained within the mine. Detector performance is also tied to persistent environmental attributes (such as soil type, terrain, vegetation, and clutter density) and transient atmospheric conditions (wind, humidity, soil moisture, and radio frequency or acoustic interference). Finally, such human factors as individual operator tuning of the detector and interpretation of the signals introduce additional sources of variability. Thus, results of reported field tests cannot be generalized, with operating characteristics being conditional on the set of experimental and field test conditions. As explained in Chapter One, such variability is clearly illustrated in the results of the International Pilot Project for Technology Cooperation field tests, shown in Figures 1.6 and 1.7. The primary implication is that even if published field results were available for all current technologies, comparing performance on these results alone would not be valid because of disparities in the testing conditions and because of variability of detector performance in different environments.

Because of the barriers to credible quantitative comparisons of innovative mine detection technologies, objective expert judgment is required to interpret the available data on detector performance. Without such judgment, it is easy to be misled. For example, many of the existing field tests have been conducted in environments in which confounding factors were minimized, yielding high probabilities of detection and low false alarm rates. Nonpartisan judgment thus was critical to the evaluations presented and to the stated tasks of the study.

To make the judgments about detector performance, we identified two sets of prominent academic, government, and private-sector scientists with expertise in landmine detection: (1) generalists, with broad and long-standing experience in the landmine detection community, and (2) specialists, currently at the forefront of developing specific cutting-edge detection technologies. Working with OSTP, we appointed eight generalists to form the RAND S&TPI/OSTP Landmine Detection Technology Task Force. Based on input from the task force members and our own independent literature review, we identified 23 leading specialists in current detection R&D efforts. Each specialist then wrote a technical paper addressing the mode of action, current capabilities and limitations, and potential future performance of his or her specific technology. We requested that each paper provide the following information:

- brief description of the basic physical principles and mine features (e.g., shape, explosives composition, metal content) exploited by the technology;
- state of development of the technology (laboratory, bench, field);
- current capabilities and operating characteristics (e.g., investigation times and examples of ROC curves from laboratory or field tests);
- known or suspected limitations or restrictions on applicability (background clutter, mine type, environmental conditions, etc.);
- estimated potential for improvement in the next two to seven years;
- outline of a sensible R&D program that could realize this potential, with rough projected costs to the extent possible; and
- references to key technical papers describing any testing (especially field testing) of the technology.

The papers are published as technical appendixes (A–W) to this report.

The appendixes provided a structure for the task force to evaluate the potential of each detection method. To conduct the evaluations, the task force held several conference calls and met for two days at RAND's Arlington, Va., office in May 2002. The "summary evaluation" section included with each technology description below indicates the results of the task force evaluations.

## INNOVATIVE ELECTROMAGNETIC DETECTION SYSTEMS

A number of innovative methods are being explored that search for buried mines based on changes in the electromagnetic properties of the surface soil and shallow subsurface. These methods include ground-penetrating radar (GPR), electrical impedance tomography (EIT), x-ray backscatter, and infrared/hyperspectral systems.

### Ground-Penetrating Radar

**Description.** GPR detects buried objects by emitting radio waves into the ground and then analyzing the return signals generated by reflections of the waves at the boundaries of materials with different indexes of refraction caused by differences in electrical properties. Generally, reflections occur at discontinuities in the dielectric constant, such as at the boundary between soil and a landmine or between soil and a large rock. A GPR system consists of an antenna or series of antennas that emit the waves and then pick up the return signal. A small computerized signal-processing system interprets the return signal to determine the object's shape and position. The result is a visual image of the object (see, for example, Figure 2.1) or an audio signal indicating that its shape resembles a landmine, based on comparison with a mine reference library.

The major design control in a GPR system is the frequency of the radio wave. The scale at which GPR can detect objects is proportional to the wavelength of the input signal, so the quality of the image improves as the wavelength decreases and the frequency increases. However, at high frequencies, penetration of the incident wave into the soil can be poor. As a result, the designer must make a tradeoff between quality of the image and required penetration depth. The optimal design for maximizing image quality while ensuring sufficient penetration depth changes with environmental conditions, soil type, mine size, and mine position. Various alternative GPR designs are being explored to optimize the tradeoff between penetration depth and image quality under a wide range of conditions. Also criti-

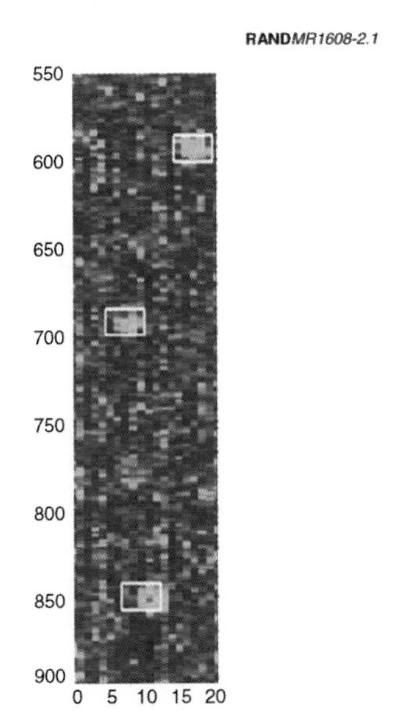


SOURCE: Ackenhusen et al. (2001).
NOTE: The bottom target is a metallic mine; the top two targets are low-metal mines.

**Figure 2.1—Images of Landmines Produced by GPR System**

cal in the design of a GPR system are signal-processing algorithms, which filter out clutter signals and select objects to be declared as mines.

GPR is a mature technology, but it has not yet been widely deployed for mine detection. GPR was first used in 1929 to measure the depth of an Austrian glacier (Olhoeft, 2002). The Army tested rudimentary GPR techniques for mine detection in the 1940s. The first commercial GPR systems were developed in 1972. Since then, use of GPR for locating buried objects ranging from utility pipelines to archaeological artifacts has proliferated. Although GPR is well established for these other uses, understanding how different environmental factors and mine characteristics affect its performance is far from complete. Until very recently, GPR was unable to meet performance targets for landmine detection established for military countermine operations (see Appendix E).

The Army is currently completing development of a landmine detection system that combines GPR and EMI. This system is called the Handheld Standoff Mine Detection System (HSTAMIDS) (see Figure 2.2). Field tested at Fort Leonard Wood, Mo., HSTAMIDS has achieved probabilities of detection of 1.00 (with a 90-percent confidence interval of 1.00–0.97) for metal antitank mines and 0.95 (with a 90-percent confidence interval of 0.97–0.93) for low-metal antipersonnel mines, with an average false alarm rate of 0.23 per square meter (see Appendix F). These test results cannot be extrapolated to predict performance in other mined environments or even under weather conditions different from those present on the day the test was performed because of the significant effects of soil type and moisture on GPR performance, as well as the variations in natural clutter objects. Nonetheless, they do illustrate the potential for a combined GPR/EMI system to achieve very high levels of performance if the integrated sensor system can be optimized to account for the effects of the local soil environment.

RAND*MR1608-2.2*

SOURCE: Minelab Countermine Division, www.countermine.minelab.com/countermine.asp.

**Figure 2.2—Prototype HSTAMIDS System**

**Strengths.** GPR has a number of advantages. First, it is complementary to conventional metal detectors. Rather than cueing exclusively off the presence of metal, it senses changes in the dielectric constant and therefore can find mines with a wide variety of types of casing (not just those with metal). Generating an image of the mine or another buried object based on dielectric constant variations is often possible because the required radar wavelength is generally smaller than most mines at frequencies that still have reasonable penetration depth. Second, GPR is a mature technology, with a long performance history from other applications. A fielded Army system (HSTAMIDS) combining GPR with EMI for mine detection already exists and is scheduled for production in 2003. Finally, GPR can be made lightweight and easy to operate, and it scans at a rate comparable to that of an EMI system.

**Limitations.** Natural subsurface inhomogeneities (such as roots, rocks, and water pockets) can cause the GPR to register return signals that resemble those of landmines and thus are a source of false alarms. In addition, GPR performance can be highly sensitive to complex interactions among mine metal content, interrogation frequency, soil moisture profiles, and the smoothness of the ground surface boundary. For example, Koh (1998) reports that because excessive water causes rapid attenuation of radio waves, GPR will perform poorly in wet soils for landmines buried below a depth of about 4 cm. However, theoretical investigations by Rappaport et al. (1999) indicate that increased soil moisture and interrogation frequency may actually strengthen the return signal for nonmetallic mines, but nonuniform soil moisture profiles (e.g., a wet surface and dry subsurface) and rough ground surfaces present difficulties. For the same mine, a given GPR can be very effective or ineffective, depending on soil moisture and mine location; such complex interplays make performance highly variable and difficult to predict. An additional limitation is that unless the GPR system is tuned to a sufficiently high frequency, it will miss very small plastic mines buried at shallow depths because the signal "bounce" at the ground surface (caused by the electrical property differences between air and soil) will mask the return signal from the mine. Finally, the GPR system designer must make a tradeoff between resolution of the return signal and depth, because high-frequency signals yield the best resolution but do not penetrate to depth.

**Summary Evaluation.** Current-generation GPR technology, such as that embodied in HSTAMIDS, has the potential for high performance. In addition, alternative approaches to GPR design (such as the Wichmann system referred to in Appendix E) have the potential to yield significant advancements over the available systems. However, the ability to model the radar response from different kinds of landmines and natural clutter is essential for yielding the expected performance gains. So far, such modeling is in its infancy. Ideally, GPR systems would be able to provide high-resolution images to a signal-processing system that could decide whether a buried object is a root, rock, clutter object, or landmine. Ralston et al. (in Appendix F) suggest development of a "library" of clutter signatures to aid in this task.

## Electrical Impedance Tomography

**Description.** EIT uses electrical currents to image the conductivity distribution of the medium under investigation. Current implementations use a two-dimensional array of electrodes placed on the ground, collecting conductivity data from stimulation of pairwise combinations of electrodes. The data are then post-processed with an algorithm that renders an image of the conductivity profile of the subsurface volume. Both metal and nonmetal mines create anomalies in the conductivity distribution that produce images, providing information about the presence and location of mines.

**Strengths.** Because both metallic and nonmetallic mines create conductivity anomalies, the technology is appropriate for detecting all types of mines. Moreover, it is especially well suited for mine detection in wet environments, such as beaches or marshes, because of the enhanced conductivity of the moist substrate. The equipment is relatively simple and inexpensive.

**Limitations.** The primary limitations are that the technology requires physical contact with the ground, which might detonate a mine, and that it cannot be used in such excessively dry, nonconductive environments as desert or rocky surfaces. The technology is also sensitive to electrical noise. Performance deteriorates substantially with the depth of the object being detected for fixed electrode array size, generally making it appropriate only for shallowly buried objects.

The resolution is not as fine as that provided by other imaging techniques, such as GPR.

**Summary Evaluation.** Because of its phenomenological limitations, EIT is probably not well suited for broad humanitarian demining needs. It has a potential niche for detecting nonmetallic mines in wet environments, a task that confounds other technologies. Even for this use, however, EIT is limited because it cannot detect at depth, and often mines in moist environments, such as rice paddies, are much deeper than those in shallow environments. A unique role for EIT in humanitarian demining is not apparent from the available information.

## X-Ray Backscatter

**Description.** Traditional x-ray radiography produces an image of an object by passing photons through the object. X rays have a very small wavelength with respect to mine sizes, so in principle they could produce high-quality images of mines. Although pass-through x-ray imaging of the subsurface is physically impossible, the backscatter of x rays may still be used to provide information about buried, irradiated objects. X-ray backscatter exploits the fact that mines and soils have slightly different mass densities and effective atomic numbers that differ by a factor of about two.

There are two basic approaches to using backscattered x rays to create images of buried mines. Methods that collimate (i.e., align) the x rays employ focused beams and collimated detectors to form an image. The collimation process increases size and weight and dramatically reduces the number of photons available for imaging. Thus, high-power x-ray generators must be used as sources. The large size, weight, and power requirements of such systems are not amenable to person-portable detectors. Alternatively, uncollimated methods illuminate a broad area with x rays and then use a spatial filter to deconvolve the system response. They may be suitable for person-portable detection.

**Strengths.** To readily distinguish mines from soils, it is necessary to use low-energy incident photons (60–200 keV). In this energy range, cross sections are roughly 10 or more times larger than is possible with most other nuclear reactions that would be applicable to mine detection. In addition, because of the reduced shielding thickness needed to stop low-energy photons, uncollimated systems can be made small and relatively lightweight. Largely because of the medical imaging industry, compact x-ray generators are now obtainable. Low-energy isotopic sources have been readily available for a long time. Practical imaging detectors are becoming more widespread, although it may be necessary to custom build for mine detection purposes. The medical imaging industry is likely to drive further advances in x-ray imaging hardware.

**Limitations.** In the required energy range, soil penetration of x-ray backscatter devices is poor. This limits detection to shallow mines (less than 10 cm deep). If source strengths are kept low enough to be safe for a person-portable system, the time required to obtain an image may be impractically long. In addition, the technology is sensitive to source/detector standoff variations and ground-surface fluctuations. Further, to image antipersonnel mines, high spatial resolution (on the order of 1 cm) is required. This may be difficult to achieve in the field. Finally, the technology emits radiation and thus will meet resistance to use because of actual or perceived risks.

**Summary Evaluation.** X-ray detection using the uncollimated imaging approach may be useful for handheld confirmatory detection of antipersonnel landmines. In fielded systems, images of mines are likely to be fuzzy but should still allow mines to be distinguished from most diffuse or elongated false alarms. On the whole, however, x-ray backscatter does not offer particular innovations or likely avenues of improvement relative to other technologies and is unlikely to yield substantial improvement in detection capabilities.

## Infrared/Hyperspectral Systems

**Description.** Infrared/hyperspectral methods detect anomalous variations in electromagnetic radiation reflected or emitted by either surface mines or the soil and vegetation immediately above buried mines (see Figure 2.3). The category encompasses technologies of diverse modes of action, including active and passive irradiation using a broad range of electromagnetic wavelengths.

Thermal detection methods exploit diurnal variations in temperatures of areas near mines relative to surrounding areas. For example,

RANDMR1608-2.3

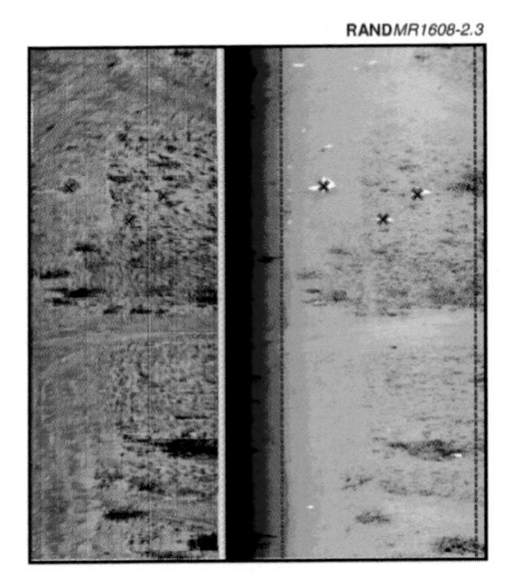

SOURCE: Space Computer Corp., www.spacecomputer.com.
NOTE: The left pane shows the infrared image, while the right pane shows the visible image. Mine locations are denoted with "x."

**Figure 2.3—Infrared Image of Mines**

mines or the soil above them tend to be warmer than surrounding areas during the day but lose heat more quickly at night. Laser illumination or high-powered microwave radiation can be used to induce these differential temperature profiles.

Nonthermal detection methods rely on the fact that areas near mines reflect light (either natural or artificial) differently than surrounding areas. Anthropogenic materials tend to preserve polarization because of their characteristically smooth surfaces, allowing discernment of surface mines. Moreover, the physical activity of emplacing mines changes the natural soil particle distribution by bringing small particles to the surface, which in turn affects the way in which the soil scatters light. Systematic changes in vegetation moisture levels immediately above buried mines also may be leveraged.

**Strengths.** These methods are attractive because they do not involve physical contact and can be used from a safe standoff distance. They are lightweight and are effective at scanning wide areas relatively quickly. When deployed from airborne platforms, they are particularly effective for detecting surface mines. Collecting and processing the signals temporally (as opposed to in "snapshots") tends to improve performance by tracking diurnal cycles.

**Limitations.** The methods, particularly thermal imaging, have been used in several prototype multisensor systems, but extreme variability in performance as a function of dynamic environmental characteristics has precluded their use for close-in detection and accurate identification of mine locations. Despite maturity of the sensor, the algorithms to process the signals in an informative way are relatively undeveloped and are not linked to physical phenomena. Thermal signatures currently are not well understood, and a comprehensive predictive model does not exist. Moreover, waves at the frequencies used by the methods cannot penetrate soil surfaces, and the localized hyperspectral anomalies produced by mine emplacement are ephemeral and are quickly eliminated by weathering. Thus, the technologies are able to detect buried mines under only limited transient conditions.

**Summary Evaluation.** With the possible exception of methods that would simulate solar heating as a means to enhance the thermal signatures of buried targets, infrared/hyperspectral methods are not particularly suitable for close-in buried mine detection. The underlying phenomena are not sufficiently characterized, and natural processes quickly erase the detectable surface anomalies. The technology has demonstrated ability and expected future promise for airborne minefield detection, especially for surface mines, but it is not expected to be useful for close-in detection of buried mines.

## ACOUSTIC/SEISMIC SYSTEMS

Acoustic/seismic methods look for mines by "vibrating" them with sound or seismic waves that are introduced into the ground. This process is analogous to tapping on a wall to search for wooden studs: materials with different properties vibrate differently when exposed to sound waves. These methods are unique among detection methods that identify the mine casing and components in that they are not based on electromagnetic properties.

## Description

Acoustic/seismic mine detection systems typically generate sound (above ground) from an off-the-shelf loudspeaker, although there are many possible configurations. Some of the acoustic energy reflects off the ground surface, but the rest penetrates the ground in the form of waves that propagate through the soil. When an object such as a mine is buried, some of the energy reflects upward toward the ground surface, causing vibration at the surface (see Figure 2.4). Specialized sensors can detect these vibrations without contacting the ground. A variety of different kinds of sensors (laser Doppler vibrometers, radars, ultrasonic devices, microphones) have been tried.

Researchers have field tested acoustic/seismic methods for landmine detection on approximately 300 buried antitank and antipersonnel

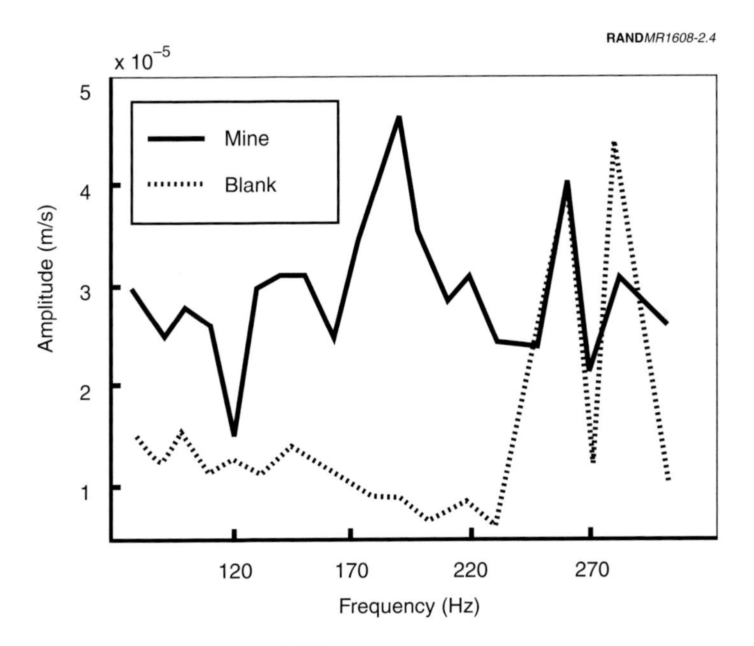


SOURCE: Rosen et al. (2000).

**Figure 2.4—Amplitude of Surface Vibration of Ground over a Mine (solid line) and a Blank (dashed line) in Response to Sound Waves**

mines and several hundred square meters of clutter locations at Army field sites in Virginia and Arizona (see Appendix G). Initial tests focused on antitank mines and yielded high probabilities of detection and low false alarm rates. For example, in one test, the acoustic system identified 18 of 19 mines buried in dirt and gravel, yielding a probability of detection of 95 percent. There was only one false alarm in the test, even though the test site was seeded with clutter items that had confounded a GPR system (Rosen et al., 2000). When the system was modified with advanced signal-processing algorithms, the false alarm rate dropped to zero.

## Strengths

Acoustic/seismic sensors are based on completely different physical effects than any other sensor. For example, they sense differences in mechanical properties of the mine and soil, while GPR and EMI sensors detect differences in electromagnetic properties. Thus, acoustic/seismic sensors would complement existing sensors well.

Acoustic/seismic systems also have the potential for very low false alarm rates. In experiments to date, false alarms from naturally occurring clutter, such as rocks and scrap metal, have been extremely low (although such hollow clutter items as soda bottles and cans would cause false alarms because the resonance patterns of these objects are similar to those of mines). An additional strength is that, unlike GPR systems, these sensors are unaffected by moisture and weather, although frozen ground may limit the sensor's capability.

## Limitations

The greatest limitation of acoustic/seismic systems is that they do not detect mines at depth because the resonant response attenuates significantly with depth. With current experimental systems, mines deeper than approximately one mine diameter are difficult to find.

Also problematic is the slow speed of existing systems. Speed currently is limited by the displacement sensor, which senses the vibrations at the surface caused by the sound waves. These displacements are very small (less than 1 $\mu$m) and are thus difficult to measure quickly in the adverse conditions of a minefield. The required scan time for locating antipersonnel mines may range from 125 to 1,000 seconds per square meter (see Appendix H). However, a number of methods are being investigated to speed up the detection process. For example, an array of N sensors will speed the system by a factor of N. Small prototypes of such arrays have been developed and can be expanded and improved with further work.

An additional limitation of existing systems is that moderate to heavy vegetation can interfere with the laser Doppler vibrometers that are commonly used to sense the vibrations at the ground surface. A new type of sensor could be developed, however, to overcome this flaw.

### Summary Evaluation

Significant progress has been made in the past five to ten years in developing acoustic/seismic mine detection systems. Interactions between the seismic waves and buried mines and clutter are much better understood, as are the seismic sources and displacement sensors. The systems show great potential, but more research is needed to make them practical. The development of an array of displacement sensors that is fast, can penetrate vegetation, and can function in the adverse conditions of a real minefield would be especially useful.

## EXPLOSIVE VAPOR DETECTION TECHNIQUES

Each detection technology discussed above searches for the casing or mechanical components of a mine. Additional research is taking place to develop methods—both biological and chemical—that identify the presence of explosive vapors emanating from mines.

Ideally, such sensors would determine whether explosive vapors are present above an anomaly located by a metal detector or other device. Although each method has a different theoretical basis, all are designed to sense low concentrations of explosive compounds or their derivatives in soil or the boundary layer of air at the soil surface. Determining the performance potential of each chemical- or biological-sensing technology requires an understanding of how explosives migrate away from landmines as well as knowledge of the chemical and physical principles of the sensor.

When a mine is buried in the soil, it almost always will gradually release explosives or chemical derivatives to the surrounding soil through either leakage from cracks and seams or vapor transport through the mine casing (in the case of plastic mines). While typically about 95 percent of the explosive will adsorb to the surrounding soil, the remaining 5 percent will travel away from the mine, mostly through dissolution in water in the soil pores (see Appendix Q). Some of this explosive will migrate to the ground surface in vapor form.

One of the key issues in detecting explosive vapors and residues is that the concentrations available for detection are extremely low (see Appendix Q). Thus the sensor must be able to operate at a very low detection threshold. The analyte that is the focus of most explosive detection research is 2,4-dinitrotoluene (2,4-DNT), which is a byproduct of trinitrotoluene (TNT) manufacturing that is present as an impurity in military-grade TNT. TNT has a very short half-life in soil (about a day at 22°C) because it is easily biodegraded and has very low vapor pressure; 2,4-DNT is much less easily biodegraded and has a higher vapor pressure, so it is the dominant chemical present in the explosive signature from most landmines. In experiments from a landmine test site reported in Appendix Q, the 2,4-DNT and 2,4,6-TNT concentrations in air above the soil were $200 \times 10^{-15}$ g per milliliter and $1 \times 10^{-15}$ g per milliliter, respectively.

Figure 2.5 summarizes the ranges of concentrations of 2,4-DNT and TNT vapors likely to be found in surface soils above landmines. To be effective, an explosive vapor detection system must be sensitive to concentrations as low as $10^{-18}$ g per milliliter if the soil is very dry or as low as $10^{-15}$ g per milliliter if the soil is moist.

## Biological Methods

Biological detection methods involve the use of mammals, insects, or microorganisms to detect explosives. Like chemical sensors, these methods rely on detection of explosive compounds rather than on detection of metal or changes in the physical properties of the subsurface. Thus, they have the potential for reducing false alarm rates from metal clutter. Each of the different methods operates on a different set of principles and is at a different stage of development. The oldest involves using trained dogs, which were first shown to be

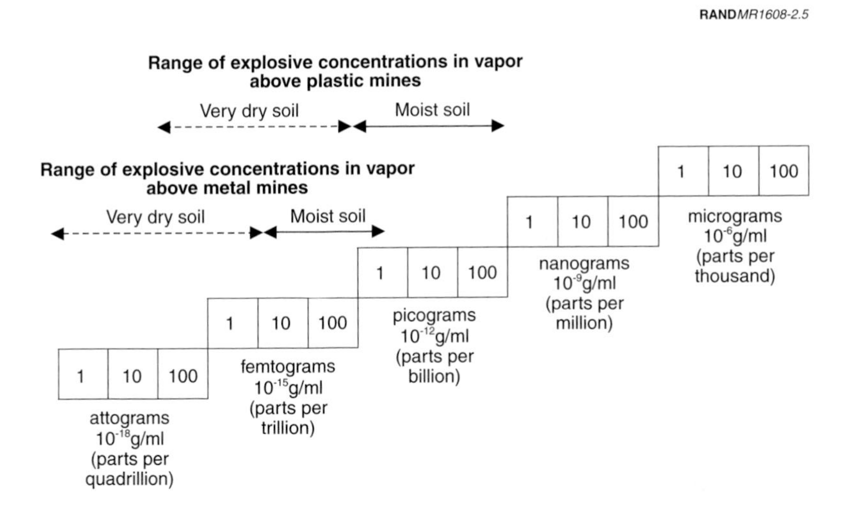


SOURCE: Adapted from a figure by R. Harmon, Army Research Office.

NOTE: To be effective in all environments, a vapor detector ideally would sense concentrations as low as $10^{-18}$ g per milliliter in dry soil and $10^{-15}$ g per milliliter in moist soil.

**Figure 2.5—Range of Concentrations of 2,4-DNT and TNT in Boundary Layer of Air Near Soil Above Landmines**

capable of smelling landmines in the late 1970s (Johnston et al., 1998). Methods employing insects and microorganisms are newer approaches that have not yet been fielded.

**Dogs and Rats.** *Description.* Mine dog detection teams have long assisted in humanitarian demining efforts. For example, more than 200 mine detection dogs currently are at work in Afghanistan. These dogs can detect mines about 95 percent of the time under favorable weather and soil moisture conditions (Horowitz et al., 1996).

Dogs have a keen sense of smell, originating from their ancestral survival needs to find food, determine territorial boundaries, and sense the presence of enemies (see Appendix T). By offering dogs a reward of food or play, they can be trained to signal when they smell mines. In the mine detection context, dogs walk ahead of their handlers, noses to the ground, and sit at the first scent of a mine (see Figure 2.6). A manual deminer then follows and investigates the area with a probe. In another application, known as Remote Explosive

Scent Tracing mode, dogs sniff at filters that have collected vapors near suspected mine locations. If a dog identifies a filter as containing explosives, then a deminer returns to the location from which the vapor was sampled to look for a mine.

Currently, dogs are capable of detecting explosive vapors at concentrations lower than those measurable by the best chemical sensors, so the lower limit at which they can detect explosives is uncertain (Phelan and Barnett, 2002). One recent study recognized that available laboratory chemical analytical methods are far from the sensitivity limits of the dog. Nevertheless, it attempted to determine the detection threshold for dogs by diluting soil contaminated with explosives to varying levels, two of which were 10 and 100 times lower (based on extrapolation, not detection) than the current chemical detection limit (Phelan and Barnett, 2002). The researchers tested the ability of three different teams of trained dogs (one from the United States, one from Angola, and another from Norway) to identify explosives in samples from the various dilutions. They found that a few of the dogs could correctly identify samples containing an estimated $10^{-16}$ g per milliliter of TNT or DNT. However, performance varied by many orders of magnitude depending on the individual dog, how it was trained, and the manner in which the training

RANDMR1608-2.6

SOURCE: RONCO Consulting Corp., www.roncoconsulting.com.

**Figure 2.6—Mine-Detecting Dog in Bosnia**

was reinforced. Further, detection performance of the dogs used in this study also appears to have been influenced by environmental conditions associated with the testing location and procedures followed, including the inadvertent use of TNT-contaminated soil samples as "clean" controls in testing at least one group of dogs.

As an alternative to using dogs or in conjunction with using dogs, researchers at the University of Antwerp have trained African giant pouch rats to detect mines. The rats are trained using food rewards to signal the presence of explosives by scratching the ground surface with their feet. Field tests of the use of rats in mine detection have begun.

*Strengths.* Canines are proven to work exceptionally well in many scenarios and under many environmental conditions. The olfactory sensitivity of some, but not all, dogs is higher than the best currently available mechanical detection methods. Advantages of using rats include the possibility that they could be deployed in large numbers and that they do not weigh enough to trigger mines, which reduces the possibility of injury.

*Limitations.* Dog performance varies widely depending on the individual dog, how it was trained, and the capabilities of the handler. Further, dogs may need to be retrained periodically because they can become confused if they discover behaviors other than explosives detection that lead to a reward. An additional limitation is that when trained to detect high levels of explosives, dogs may not automatically detect much lower levels and may need to be specially trained for this purpose. Like other methods that rely on vapor detection, performance of mine detection dogs can be confounded by environmental or weather conditions that cause explosive vapors to migrate away from the mine or that result in concentrations of vapors that are too low even for dogs to detect. Rats likely would have similar limitations.

*Summary Evaluation.* Canines are proven performers and a valuable asset in demining. However, continued investigation of the sensitivity of canine olfaction and how this varies with the dog and with training is necessary to understand the factors that affect reliability. Additionally, the vapor and particle signature of the mine in the field must continue to be investigated to better understand performance potential for canines. Additional research to explore the potential for deployment of African giant pouch rats in demining also is warranted.

**Bees.** *Description.* By lacing sugar with a target chemical and placing the sugar in the bees' natural foraging area, bees can be trained to associate the chemical odor with food and to swarm over any location containing the target odorant. Entomologists have trained bees to detect a variety of explosives and have been researching ways to use trained bees in humanitarian demining. There are two suggested strategies. The first involves monitoring the movement of bees trained to detect explosives and keeping track of the locations where they swarm. The second involves sampling the beehive for the presence of explosives, which can be transported to the hive on the bees' mop-like hairs.

Several field tests have been conducted to investigate the potential use of bees in mine detection (see Appendix S). The most recent and comprehensive test involved placing DNT in petri dishes, covering the DNT with sand, and placing the dishes in a flat, open space for subsequent detection by the bees. In these experiments, bees proved capable of detecting DNT concentrations that were estimated to be 0.7–13.0 ppb (approximately $10^{-12}$ g per milliliter). Earlier lab testing had indicated that bees could detect concentrations down to 20 ppt.

*Strengths.* Trained bees detect explosives and therefore are not limited by the same types of false alarms that plague metal detectors. They also potentially could search a relatively large area in a short time.

*Limitations.* As for chemical and bacterial detection systems, more needs to be understood about the fate and transport of explosives in the subsurface before the full potential of trained bees to detect landmines can be understood. To date, no field trials using actual mines have been conducted. Further, bees can only work under limited environmental and weather conditions. They do not work at all at temperatures below 40°F. In addition, all tests to date have been in clear, open fields; whether bees would perform in forested or other heavily vegetated environments is unknown. The inability to track bee movements also currently poses difficulties.

*Summary Evaluation.* Continued investigation of the use of bees in mine detection is warranted. Experiments under more realistic conditions could give a better indication of the potential of this method. However, clear decision points should be established for determining whether to continue with research funding, if the method continues to look promising, or to terminate it, if there are insurmountable obstacles.

**Bacteria.** *Description.* During the 1990s, researchers engineered a strain of bacteria that fluoresce in the presence of TNT (see Appendix R and Kercel et al., 1997). A regulatory protein in the bacteria recognizes the shape of the TNT molecule and fluoresces whenever the molecule is present (see Figure 2.7). In principle, a bacterial mine detection process would involve spraying bacteria on the mine-affected area, possibly using an airborne system. The bacteria would be allowed to grow for several hours. Then, a survey team would return to search for fluorescent signals. The search could be conducted either from an airborne system or using a handheld fluorescence detector.

One field trial using bacteria has been conducted. Five targets containing from 4 oz to 10 lb (100 g to 5 kg) of TNT were placed in a

RANDMR1608-2.7

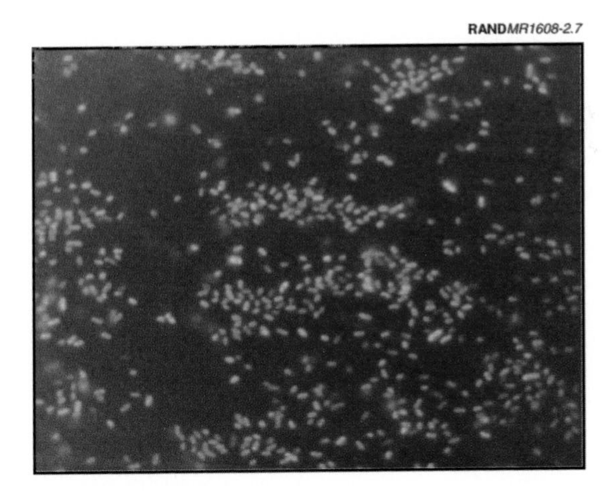

SOURCE: http://attic.jcte.jcs.mil/documents/forerunner_07_01.pdf.

**Figure 2.7—Bacteria Fluorescing in the Presence of TNT**

quarter-acre field site. The bacteria detected all five targets, but there were also two false alarms. Based on this single field trial, it is not possible to determine the lowest concentration of explosive that bacteria are capable of detecting.

*Strengths.* Like chemical sensors, bacteria can be engineered to be highly specific to the explosive of concern. The regulatory protein that causes the bacteria to fluoresce recognizes only TNT and structurally similar molecules. Thus, this method has the potential to reduce false alarms from clutter objects. An additional advantage is that it may allow coverage of a large area in a relatively short time. In theory, the unit cost of this method should decrease as the size of the search area increases.

*Limitations.* The limited research to date has revealed possible environmental limitations of this method. Bacteria are highly sensitive to environmental conditions. The existing strain used to locate TNT cannot survive at extreme temperatures. In addition, the method functions only in moist soil because dry soil quickly absorbs the bacteria. Another limitation is that the potential for false alarms is unknown. For example, the two false alarms in the single field test could have resulted from the migration of explosives away from the targets or from some other chemical in the environment that triggered the fluorescent response. An additional problem in experimental trials was that the fluorescence detector missed some of the signals from the bacteria. Finally, the performance potential of this method will be limited by the fate and transport of explosives in the subsurface. If the explosives migrate away from the mine, then the bacterial signal may occur at a distance from the mine. In addition to these operational limitations, public concerns about introducing genetically engineered organisms into the environment may limit the application of bacteria in mine detection.

*Summary Evaluation.* The potential for using bacteria in mine detection remains largely untested, except for the single field trial referenced in Appendix R. Nonetheless, continued investigation is warranted as long as clear decision points for terminating or continuing funding are established. For example, if research shows that environmental confounding factors (such as moisture conditions and temperature) preclude the use of bacteria in all but an ideal environment, then research may need to be halted unless bacterial strains can be bred to overcome these limitations.

## Chemical Methods

**Description.** A variety of possible nonbiological mechanisms for detecting low concentrations of explosives in air or in soil samples have been investigated in recent years (see Table 2.1). Most of these investigations resulted from DARPA's "Dog's Nose" program, which sponsored R&D leading to the development of highly sensitive odor detection devices. Some of the techniques were patterned after the mammalian nose. For example, one approach uses arrays of polymer-based sensors that detect explosive vapors (and other volatile chemicals) based on the amount of swelling in the polymers

**Table 2.1**

**Chemical and Physical Methods for Sensing Explosive Vapors**

| Sensor Category | Description | Approximate Detection Limit (g explosive per ml air) |
|---|---|---|
| Fluorescent | Measure a change in fluorescence wavelength on the tip of a polymer-coated glass fiber or on an antibody biosensor that occurs in response to the presence of explosives | $10^{-15}$ |
| Electrochemical | Measure changes in electrical resistance of arrays of polymers upon contact with explosive vapors; alternatively, measure changes in electrical properties in coupled electrode pair during reduction or oxidation of explosives | $10^{-12}$ |
| Piezoelectric | Measure shift in resonant frequency of various materials (thin polymer, quartz microcrystal, or other) due to mass change upon exposure to explosive vapor | $10^{-11}$ |
| Spectroscopic | Compare the spectral response of a sample with that of a reference material | $10^{-9}$ |

caused by exposure to the analyte of concern (Freund and Lewis, 1995; Doleman and Lewis, 2001; Hopkins and Lewis, 2001). These swelling differences, in turn, lead to easily detected changes in electrical resistance of the polymers. Like mammalian noses, these sensors rely on low-level responses of multiple different types of receptors (many polymer types in the case of the mechanical sensors, and many types of receptor proteins in the case of the mammalian nose).

Of the various vapor sensors, a system using novel fluorescent polymers is closest to being deployable and currently has the lowest detection limit, as shown in Table 2.1 (Yang and Swager, 1998; Cumming et al., 2001). The sensor consists of two glass slides, each covered by a thin film of the fluorescent polymer. When a sample of air containing explosives passes between the slides, some of the explosive binds to the polymer and in the process temporarily reduces the amount of fluorescent light that the slides emit. A small photomultiplier device detects the reduced light emission, and electronics signal the operator that explosives are present. Nomadics Inc. has developed a man-portable prototype of this system and has conducted limited field tests (see Figure 2.8). The current system can

RAND*MR1608-2.8*

SOURCE: Nomadics Inc., www.nomadics.com.

**Figure 2.8—Nomadics' Sensor for Detecting Explosive Vapors**

detect explosive vapor concentrations as low as $10^{-15}$ g per milliliter, and additional development work is expected to lower this threshold (Cumming et al., 2001).

**Advantages.** Vapor and residue sensors detect explosives and therefore could serve as complements to detection devices that rely on physical features of the mine. In addition, most of the methods have the potential to be engineered as small, lightweight, easily transportable, and simple-to-operate systems with relatively low power requirements. The Nomadics prototype already available is comparable in size to a typical metal detector and, like a metal detector, can operate at a walking pace. It has an extremely low detection threshold ($10^{-15}$ g per milliliter).

**Limitations.** Perhaps the greatest obstacle that these systems must overcome is the need for a probability of detection of one if they are to replace manual prodders for confirming the presence of mines. No deminer will exchange a conventional prodder for a device that has less than near-perfect probability of detection because of the obvious safety concerns involved. The detection sensitivity of current technologies, with the exception of the fluorescent polymer approach, is not low enough to provide for reliable detection of metal-encased mines in dry soil, and even this method may not perform well in the driest of environments. Another problem is that the presence of explosive residues in soil from sources other than landmines will trigger false alarms. This limitation would apply primarily to battlefield environments because dispersed explosives remaining after weapons are fired will biodegrade over time. Naturally occurring chemicals that react with the polymers also may cause false alarms. For all of these systems, the effects of variations in environmental conditions, especially soil moisture, on performance is not well understood. Further, the location at which explosive vapors are present at the highest concentration is often displaced from the mine location. Current understanding of explosives fate and transport from buried mines is insufficient to allow for the reliable location of a mine based on measurement of the extended explosive vapor signal.

**Summary Evaluation.** Explosive vapor and residue detectors have the potential to be used as confirmatory sensors for landmines if the probability of detection can be increased to near one. Whether this is possible cannot be determined without additional basic research.

Research is necessary to establish the lower limits of detection for the different types of vapor sensors. Also needed are further investigations to allow quantitative modeling of the amounts and locations of explosives available for detection at the surface under different environmental conditions.

## BULK EXPLOSIVE DETECTION TECHNIQUES

Biological and chemical methods for detecting explosive vapors currently are limited by incomplete knowledge of how explosive vapors migrate in the shallow subsurface. An additional category of explosive detection technologies overcomes this limitation by searching for the bulk explosive inside the mine. Methods being explored for this purpose include nuclear quadrupole resonance (NQR) and a variety of methods that use the interaction of neutrons with components of the explosive. These technologies emerged from interest in detecting bulk explosives in passenger baggage for the airline industry and in investigating the potential presence of explosive devices in other settings.

### Nuclear Quadrupole Resonance

**Description.** NQR is a radio frequency (RF) technique that can be used to interrogate and detect specific chemical compounds, including explosives. An NQR device induces an RF pulse of an appropriate frequency in the subsurface via a coil suspended above ground (see the prototype in Figure 2.9). This RF pulse causes the explosives' nuclei to resonate and induce an electric potential in a receiver coil. This phenomenon is similar to that exploited by magnetic resonance imaging (MRI) used in medical testing, but NQR uses the internal electric field gradient of the crystalline material rather than an external static magnetic field to initially align the nuclei.

**Strengths.** NQR has a number of features that make it particularly well suited for landmine detection. The primary attraction of NQR is its specificity to landmines: In principle, it signals only in the presence of bulk quantities of specific explosives. Unlike many other technologies, its false alarm rate is not driven by ground clutter but rather by its signal-to-noise ratio (SNR). The SNR increases with the

RANDMR1608-2.9

SOURCE: A. Hibbs, Quantum Applied Science and Research Inc.

**Figure 2.9—Prototype Handheld NQR Mine Detector**

square root of the interrogation time and also increases linearly with the mass of the explosive. Thus, with sufficient interrogation time, NQR can achieve nearly perfect operating characteristics (probability of detection near one with probability of false alarm near zero). This makes NQR more attractive as a confirmation sensor used to interrogate only those locations identified by other detectors (e.g., GPR, EMI) as likely mine locations. Interrogation times of 0.5–3.0 minutes may be sufficient for performance that leads to high probability of detection (more than 0.99) and low probability of false alarm (less than 0.05). The NQR signal from cyclotrimethylenenitramine (RDX) is particularly large, implying high performance and small interrogation times (less than three seconds) for detection of mines containing RDX. Another positive feature of NQR is that it is relatively robust to diverse soil conditions; for example, because it requires bulk concentration of explosives to declare, it is not misled by trace explosive residues as can be the case with vapor-sensing techniques.

**Limitations.** The major weakness of NQR is the fact that, because of its nuclear properties, TNT, which comprises the explosive fill of most landmines, provides a substantially weaker signal than either RDX or tetryl, posing a formidable SNR problem. Moreover, TNT inherently requires longer interrogation times because its nuclear properties preclude interrogation more frequently than once per five to ten seconds. Another significant limitation is the susceptibility of NQR to RF interference from the environment. This is especially problematic for TNT detection because the frequencies required to induce a response from TNT (790–900 kHz) are in the AM radio band. When present, radio signals overwhelm the response from TNT.

An additional weakness is that NQR cannot locate explosives that are encased in metal because the RF waves will not penetrate the case. This is not a major weakness because a large majority of antipersonnel mines have plastic cases, and EMI detection can successfully detect those with metal cases. NQR also cannot detect liquid explosives, but very few antipersonnel mines use liquid explosives.

NQR is very sensitive to the distance between the detection coil and the explosive. Therefore, the detection coil must be operated very close to the ground, which can be problematic in rough or highly vegetated terrain. Moreover, current implementations require stationary detection for optimal results; detection in motion substantially degrades the SNR.

**Summary Evaluation.** NQR is an explosive-specific detection technology that offers considerable promise as a technique for reducing false alarms as compared with such conventional detection approaches as EMI. It offers opportunities for improvement not addressed by competing technologies—most notably that its potency derives from unique explosive signatures of mines. In principle, this specificity affords it the possibility of a zero false alarm rate against nonmetal mines. Currently, the most promising role of NQR is that of a confirmation sensor used in conjunction with a conventional scanning sensor or as part of an integrated multisensor detection system.

## Neutron Methods

**Description.** Neutron interrogation techniques involve distinguishing the explosives in landmines from surrounding soil materials by probing the soil with neutrons and/or detecting returning neutrons. Differences in the intensity, energy, and other characteristics of the returning radiation can be used to indicate the presence of explosives.

Only three of the many possible reactions involving neutrons or gamma rays have reasonable potential for landmine detection. The first, *thermal neutron analysis*, is the only nuclear technique that is currently fielded by a military: The Canadian military uses it as a vehicle-mounted confirmatory detector for antitank mines (see Appendix N). Size and weight limits imposed by physics preclude it from being person-portable or being able to detect small antipersonnel mines in practical applications. The second method, known as *fast neutron analysis*, has similar limitations. The third method, *neutron moderation*, is the only one of the three with the potential to yield a person-portable detector for antipersonnel mines. Neutron moderation discerns buried materials with low atomic numbers (e.g., hydrogen).

**Strengths.** The physical properties of neutron moderation allow the technology to use low-strength source radiation, which reduces shielding required to protect workers from radiation exposure. Thus, designing a handheld system may be possible. Costs of a production imager are expected to be moderate.

**Limitations.** Neutron activation methods can, at best, measure relative numbers of specific atoms but cannot determine what molecular structure is present. Because neutron moderation is most sensitive to hydrogen, hydrogenous materials, particularly water, produce many false alarms. Thus, to detect landmines successfully, it is necessary to use the response from the neutrons to generate a visual image of the area under investigation. Simulations show that the method will work in soil with 10-percent moisture or less and may be usable when moisture content is as high as 20 percent. Ground-surface fluctuations and sensor height variation also contribute to false alarms in nonimaging systems. Imaging can reduce these effects, although some degradation of the image is expected.

With sources having sufficiently low strength to be practical for handheld use, a few seconds will be required to acquire an image. This makes neutron moderation imaging more suitable for confirmation than for primary detection. Also, there is a perceived (more than actual) radiation hazard associated with nuclear techniques that must be overcome by the users. Broad-area, low-power electronic neutron sources, under development by Defence R&D Canada, could reduce this perceived risk.

**Summary Evaluation.** The majority of neutron technologies have physical limitations that preclude them from being portable. Only one technology—neutron moderation imaging—may be useful for handheld confirmation of antipersonnel landmines. In fielded systems, images of these mines are likely to appear as fuzzy blobs, but that will still allow mines to be distinguished from most diffuse or elongated false alarms generated by moisture. On balance, however, neutron moderation imaging is very unlikely to yield substantial improvements in detection speed beyond what is capable with other confirming detectors.

## INNOVATIVE PRODDERS AND PROBES

The last step in mine detection has long been probing manually to determine whether a signaled item is a mine or just harmless clutter. The probe operator learns through experience to feel or hear the difference between a mine casing and other buried objects. Probing is dangerous. Deminers often inadvertently apply sufficient force to the mine to detonate it. This excess force does not always cause the mine to explode, but when it does, the probe itself can become a deadly projectile. New concepts are being explored to improve the safety of probing and to help discriminate mines from clutter.

### Description

Research to improve prodders and probes has followed two lines of investigation: (1) development of probes that would signal to the deminer when too much force is being applied and (2) development of "smart" probes that provide information about the characteristics of the item being investigated. These latter probes are intended to provide information about some physical, electromagnetic, or chemical characteristic of the object being investigated in order to identify it. The only such probe engineered to date delivers an acoustic pulse to determine whether the object is plastic, metal, rock, or wood (see Appendix W). Performance of this smart probe was mixed in limited testing: In a Canadian test, it correctly identified 80 percent of the mines, but in a U.S. test it identified only 69 percent of the mines. An improved version identified 97 percent of the mines in a Canadian test. An alternative type of smart probe that is in the research stage sprays focused jets of cold or heated water at regular intervals to detect mines by sound or thermal signature.

## Strengths

Probing is an established step in manual demining. Improved probes could decrease the risks to deminers by providing feedback about the nature of the object being investigated. In addition, theoretically, a probe could deliver any of a number of different detection methods (acoustic, electromagnetic, thermal, chemical, etc.), and the proximity of the probe to the landmine could improve performance. For example, methods based on identifying explosive vapors likely would perform better in close proximity to the mine, where vapor concentrations are much greater than those on the surface. However, such advanced probes have yet to be developed.

## Limitations

Any improved probe must essentially identify mines 100 percent of the time to be accepted by the demining community. This is because deminers view their current conventional probes as 100-percent accurate. Some field testing has borne this out: A conventional military prodder in one field experiment correctly distinguished between 38 mines and 119 rocks (see Appendix W). In addition, instrumented probes may not be useful in dense ground or in ground with extensive root structures.

## Summary Evaluation

Because the probe is likely to remain part of the deminer's tool kit for the foreseeable future, limited efforts to develop more sophisticated probes should continue. To be useful in speeding up demining operations, probes must be able to identify rapidly and accurately whether the detected item is a mine or another object. Research should continue on instrumented probes with detection devices that analyze the material content of the item under investigation. To increase probe safety, work should continue to develop probes that indicate the level of force being applied.

## ADVANCED SIGNAL PROCESSING AND SIGNATURE MODELING

This chapter described a diverse range of innovative sensors for obtaining signals that indicate the presence of landmines. Underlying nearly all of these technologies are algorithms that translate these signals into information that can be used (by either a human operator or an automated system) to make a declaration decision. Traditionally, these algorithms have been very simple and have focused on detecting anomalies in the subsurface, as opposed to providing information about the size, shape, or chemical content of the object. For example, conventional handheld EMI detectors provide information only about the strength of the return signal. Typically, the frequency of the audible tone increases as the received signal strength increases. The operator then decides what signal strength to use as an indicator of the possible presence of a mine. Without the additional step of discriminating background clutter from legitimate targets, the number of false alarms is large and hampers the detection and clearance process.

The primary goal of advanced signal-processing algorithms is to maximize the use of information generated by the sensor to help discriminate targets from background clutter. Recent efforts have combined information from physical models specific to the particular sensor technology, statistical analyses of the generated signals, and spatial information to achieve this goal (see Appendixes U and V). Advanced algorithms leveraging statistical models have been shown to reduce the false alarm rate for EMI, GPR, and combined sensors in certain settings (Collins et al., 2002; Lewis et al., 2002; Witten et al., 2002). Because of fundamental physical limitations of the technologies, no amount of signal processing will eliminate all false alarms from EMI and GPR systems. Nonetheless, such advanced algorithms are important to improving mine detection technologies. For nearly all types of sensors, advanced algorithms could be or are being developed that make efficient use of the generated signal to improve operating characteristics. More important, advanced signal-processing and signal fusion methods are crucial to the development of next-generation mine detection systems discussed in the next chapter.

Chapter Three

# MULTISENSOR SYSTEM TO IMPROVE MINE DETECTION CAPABILITY

No single mine sensor has the potential to increase the probability of detection and decrease false alarm rates for all types of mines under the wide variety of environmental conditions in which mines exist. As is clear from the previous chapter, each innovative method is subject to limitations under certain conditions of environment and mine type. Thus it is unlikely that any one technology will provide the breakthrough necessary to substantially improve humanitarian demining operation times.

To achieve substantial decreases in mine detection time while maintaining high probabilities of detection, a paradigm shift is needed in mine detection R&D. Rather than focusing on individual technologies operating in isolation, mine detection R&D should emphasize the design from first principles and subsequent development of an integrated, multisensor system that would overcome the limitations of any single-sensor technology. The goal of such R&D should be to produce a single system comprised of many different types of sensors and algorithms for combining the feedback from all sensors in an optimal manner.

Multisensor systems that combine technologies with different sources of false alarms could substantially decrease the false alarm rate. As an example, Table 3.1 lists the individual sensor technologies that, based on the evaluations in the previous chapter, appear most promising for close-in mine detection for humanitarian operations. As shown, the methods have substantially different sources of false

**Table 3.1**

**Sources of False Alarms for Selected Mine Detection Technologies**

| Detection Technology | Primary Source of False Alarms |
| --- | --- |
| EMI | Metal scrap, natural soil conductivity, and magnetism variation |
| GPR | Natural clutter (roots, rocks, water pockets, etc.) |
| Acoustic/seismic | Hollow, man-made objects (e.g., soda cans) |
| Fluorescent polymers | Explosive residues |
| NQR | Radio frequency interference |

alarms. Multisensor systems also could increase the probability of finding diverse types of mines and of operating effectively in a range of environments. Table 3.2 shows the most promising technologies and the mine and environmental conditions for which they are best suited. As can be concluded from Tables 3.1 and 3.2, combining technologies could yield a system that is robust across diverse environments and mine types.

The potential of multisensor systems to improve detector operating characteristics has been demonstrated empirically. In a comparative field study of different sensor fusion algorithms, Gunatilaka and Baertlein (2001) showed that fusion of EMI, GPR, and infrared data achieved better than a factor of 8 improvement in the probability of false alarm compared with the best individual sensor, at PD = 1. Similarly, Collins (2000) found that advanced fusion of EMI, GPR, and magnetometry signals achieved between a factor of 5 and 12 improvement in probability of false alarm compared with simple thresholding on raw data of individual sensors, at PD = 0.8. Finally, in an extensive field test of five vehicle-mounted systems consisting of metal detectors, GPR, and infrared sensors conducted by the Institute for Defense Analyses (IDA), fusion of the three signals improved operating characteristics for all systems compared with any single sensor or pair of sensors (Rotondo et al., 1998). These studies, and others cited in the appendixes, demonstrate that a well-designed multisensor system has the potential to substantially improve performance relative to detectors that use a single-sensor technology.

**Table 3.2**

**Mine Types and Soil Moisture Conditions for Which Selected Mine Detection Technologies Are Best Suited**

| | Mine Types and Soils for Which the Technology Is Most Effective | | | | | |
|---|---|---|---|---|---|---|
| Detection Technology | Metal Mines | Low-Metal Mines | TNT Mines | RDX Mines | Dry Soil | Wet Soil |
| EMI | √ | | √ | √ | √ | √ |
| GPR | √ | √ | √ | √ | √ | |
| Acoustic/seismic | √ | √ | √ | √ | √ | √ |
| Fluorescent polymers | √ | √ | √ | √ | | √ |
| NQR | | √ | | √ | √ | √ |

## KEY DESIGN CONSIDERATIONS

A multisensor system that achieves performance superior to a single-sensor system across a diversity of environments and target types will require effective and flexible methods for combining information from multiple sensors of different modalities. Multisensor systems can combine or "fuse" the information from the component sensors in a variety of ways, characterized by the stage of signal processing at which the information is combined. Broadly speaking, the different fusion methods can be categorized as "decision-level," "feature-level," or "data-level" (Ackenhusen et al., 2001; Gunatilaka and Baertlein, 2001; Dasarathy, 1994). In decision-level fusion, each component sensor of the system provides the operator with a declaration decision from independently processed signals, and these are combined to make the overall declaration decision. For example, HSTAMIDS (see Chapter Two) produces two separate signals—one from the GPR detector and another from the EMI detector—and the operator must decide whether to investigate an item, depending on the signals received. "Hard" decision-level fusion bases the overall decision on only the individual binary decisions (declare/non-declare) of the component sensors, generally using simple rules (e.g., Boolean "and," "or," or majority voting). Alternatively, if individual declaration decisions are augmented with some measure of confidence, "soft" decision rules that give more weight to more reliable decisions are possible. In either case, the overall decision is based on only the independently processed signals from the individual sensors. This is in contrast to feature- or data-level fusion, in which the signals generated by each sensor are combined algorithmically to present the operator with a single signal on which to base the declaration decision. That is, rather than combining the processed outputs of various sensors, these lower-level fusion methods jointly process the received physical information at the signal level. Data-level fusion combines the raw data collected by each sensor, while feature-level fusion combines information about informative "features" extracted from the raw signals.

Situations occur in which one type of fusion might be advantageous, relative to another. For example, if the multiple sensors detect the same type of target but have different confounders, decision-level fusion helps to reduce errors and improve operating characteristics. However, feature-level fusion would likely be more effective for an array of sensors, each of which is designed to detect a different kind of target. Moreover, there are theoretical reasons to believe that properly implemented fusion at lower information levels (e.g., data or feature) should be superior to fusion at higher (e.g., decision) levels. This is because such fusion can make more efficient use of the available information by exploiting the simultaneous characteristics of the signals ignored by decision-level fusion. Lower-level fusion also in principle offers a pragmatic advantage because it operates as an integrated unit, presenting the user with a single signal in a manner similar to traditional EMI detectors. "Primary" and "confirmatory" sensing would be transparent to the operator. Alternatively, when multiple signals are presented to the operator, multiple operating thresholds must be chosen. This makes performance optimization more cumbersome.

Most work to date has used decision-level fusion, which has been shown to reduce false alarm rates relative to individual sensor performance while maintaining high probabilities of detection. Data- or feature-level fusion is more difficult, less mature, and thus far has been shown to offer relatively modest incremental improvements to decision-level fusion. For example, in a performance study of decision- versus feature-level signal processing on data collected from EMI, GPR, and infrared sensors, Gunatilaka and Baertlein (2001) demonstrated that hard decision-level fusion did not perform appreciably better than the best of the individual sensors (GPR in this case). However, soft decision- and feature-level fusion offered marked benefits, with the feature-level algorithm providing a modest but meaningful additional benefit relative to the soft decision-level methods.

Because feature- and data-level fusion are relatively new research areas, it is uncertain how rapidly and to what degree their theoretical advantages over decision-level fusion will result in practical applications. Thus, in designing the multisensor system we propose, we believe it is important to keep an open mind about how to process information from multiple sensors. Researchers should be receptive to combining information at whatever level is necessary to achieve optimal performance in practice. However, given the potential to improve operating characteristics and simplify multisensor system operation (by presenting a single signal), research should continue on advanced lower-level fusion algorithms.

Another formidable challenge to the next-generation multisensor system is having the flexibility to be effective across a broad range of field conditions and target types. Regardless of the level at which information from the component sensors is combined, performance using a fixed configuration of operating thresholds will vary tremendously across the diversity of field conditions encountered in humanitarian demining operations. It is impossible to determine optimal thresholds for all possible field conditions a priori. Thus, achieving optimal performance will require adjustments to operating thresholds on a case-by-case basis. Such adaptation may not be practical for some military demining operations but is practical for most humanitarian demining operations. The mode of adaptation conceptually falls along a continuum of end-user responsibility. At one extreme, the user is completely responsible for optimizing the system for the field, with no built-in system intelligence for this task. At the other extreme, the system is entirely self-calibrating and learns about the field in an automatic, real-time fashion. While the latter ideal may be a worthy long-term goal, it is unrealistic in the short term, and initial development efforts should compromise between system and user optimization. Through both field testing and aggressive environmental and target-sensor interaction modeling, it should be possible to learn what environments and targets are most challenging to the system and how the system can be altered to handle them effectively. This information can be embedded into the system logic, and the user would then input information about a small number of high-leverage field condition variables that would result in gross alterations to system thresholds. The remaining user performance tuning could then be along some manageably simple dimensions, ideally of complexity on par with current EMI detectors. We envision that, as research progresses, more of the local adaptation can be off-loaded to the system.

## POTENTIAL FOR A MULTISENSOR SYSTEM TO INCREASE MINE CLEARANCE RATE

The primary motivation for pursuing multisensor systems is to reduce the false alarm rate and in turn decrease the amount of time spent on mine detection. In requesting this study, OSTP asked that RAND S&TPI assess whether improved detection methods could increase the speed at which mines can be cleared by an order of magnitude. While we conclude that multisensor systems are the most promising path for improved performance, we also determine that it is unlikely that such systems will achieve order-of-magnitude improvements in clearance rates in the near future. There are two primary reasons for this conclusion. First, reductions in the false alarm rate do not bring about equal reductions in mine clearance time, even under the assumption that the time spent removing actual mines is negligible. Depending on the specific environment, a nontrivial portion of the total clearance time may be spent on such "overhead" operations as vegetation clearance. In some cases, these preclearance activities may require so much time that even a detection system with perfect operating characteristics could not achieve an order-of-magnitude improvement in clearance rates. For example, Treveylan estimates that an improved detection system that eliminated 99 percent of false alarms would speed overall clearance rates by 60–300 percent of the original clearance rate (meaning 1.6–4.0 mines could be cleared in the time now required to clear one mine), depending on the density of vegetation (Treveylan, 2002a,b). This is in contrast to the hundredfold (i.e., 9,900-percent incremental) improvement that is theoretically possible if only the time spent investigating false alarms is considered.

The second reason we conclude that order-of-magnitude improvements are unlikely in the near future stems from tradeoff between decreasing the false alarm rate and increasing the probability of detection. For single-sensor systems, reducing the false alarm rate decreases the probability of detection. For the near future, the same phenomenon is likely to occur even for a well-designed multisensor system. Because of the resulting increased residual risk from more missed mines, this is not acceptable. Thus, the goal of minimizing the false alarm rate will continue to be limited by the need to maximize the probability of detection.

Although order-of-magnitude decreases in mine clearance rates are unlikely to be achieved in the near term, focused investments in multisensor system development would yield substantial savings in time and money. Reductions in false alarms translate into reductions in operation times, even if these reductions are not proportional; therefore, operation costs should decrease substantially. For example, the United Nations has estimated that the average cost of clearing a mine is $300–1,000 per mine (Hubert, 1998). Based on this average, the cost to clear all 45–50 million mines worldwide will total $14–50 billion. Given that a large fraction of these costs derive from paying deminers for time spent investigating false alarms, even moderate time savings could save billions of dollars. For example, Treveylan (2002a,b) has estimated that if 50 percent of false alarms were correctly declared as nonhazardous, then demining speed would improve by approximately 30 percent in a highly vegetated country, such as Cambodia, and by approximately 60 percent in a minimally vegetated country, such as Afghanistan, compared with a detector that has no ability to distinguish false alarms from mines. If one assumes that the improved system would increase demining speed by 30–60 percent worldwide and that such gains in speed translate directly into cost reductions, then the improved detector would save 23–38 percent of the total demining cost of $14–50 billion. Improvements beyond an ability to correctly identify 50 percent of false alarms are expected, with proportionately higher decreases in mine clearance time and costs. For example, Treveylan (2002a,b) estimates that a system that correctly identified 90 percent of false alarms would speed demining by 60 percent in Cambodia and 200 percent in Afghanistan.

In addition, the development of an effective multisensor system, and the resulting reduction in false alarms, has numerous tangible benefits beyond reduction in operation times. The vast number of false alarms encountered in demining operations may foster inattention and carelessness, which may increase the occurrences of demining accidents. A reduction in the false alarm rate thus has the potential to improve demining safety. Moreover, although the discussion of multisensor systems has focused on the potential for minimizing the false alarm rate, it is possible that modest false alarm rate reductions could be coupled with improvements, rather than degradations, to the probability of detection. This ultimately improves public safety. Also, the R&D necessary to pursue the integrated multisensor system will require aggressive interdisciplinary efforts by top researchers and is likely to result in technological advances that can be leveraged in other applications of public interest (such as chemical weapon detection, airline safety, drug enforcement, military countermine operations, and improvised explosive device detection). Finally, even if order-of-magnitude improvements are not achievable in the near term, incremental improvements may lead to an order-of-magnitude improvement over time.

## CURRENT U.S. R&D INVESTMENT IN MINE DETECTION TECHNOLOGIES

The primary conclusion of this report is that development of an advanced, integrated multisensor system is the most likely pathway to substantial improvements in humanitarian demining operations. However, the federal government currently has no research program that is directed specifically at designing such a system. While the United States funds research related to humanitarian mine detection, the amount allocated is very small. Further, existing funding is not optimized toward design from first principles of an advanced multisensor system.

Currently, federal funding for R&D on humanitarian mine detection is limited. The only federal R&D program dedicated to humanitarian demining is the Department of Defense Humanitarian Demining R&D Program (HD program), established in 1995 and administered by the Assistant Secretary of Defense for Special Operations/Low-Intensity Conflict. Total funding for this program in 2002 was $13.5

million (UXO Center of Excellence, 2002). Of that amount, $4.9 million was allocated for detection technologies.[1] Nearly half ($2.2 million) of the detection technology budget was spent on wide-area detection R&D (for use in remote identification of potential mine field locations). Thus the total amount available for R&D on close-in detection technologies was $2.7 million.

Existing federal funding for humanitarian mine detection is focused primarily on optimizing the performance of EMI and GPR systems. As shown in Figure 3.1, most of the $2.7 million spent on close-in methods in 2002 was concentrated on GPR and EMI. The only other detection method included was chemical vapor–sensing technologies. No funding was invested in researching a multisensor system.

The HD program emphasizes traditional detection methods because it is an applied research program and is by design not positioned to

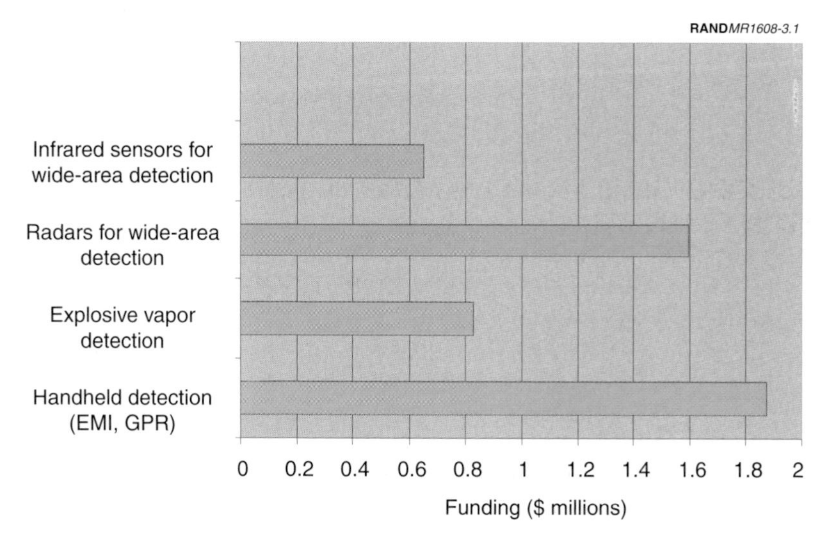


NOTE: Total funding for detection technologies was $4.9 million. Data in figure are from R. Weaver and S. Burke (personal communication, August 15, 2002).

**Figure 3.1—Detection Technologies Funded by the U.S. Humanitarian Demining R&D Program in Fiscal Year 2002**

---

[1]Information obtained from personal communication with R. Weaver and S. Burke, Army Night Vision and Electronic Sensors Directorate, Ft. Belvoir, Va., August 13, 2002.

conduct basic research. The U.S. Department of Defense organizes its research into five tracks corresponding to level of maturity of the technology, with track 6.1 dedicated to basic research projects and tracks 6.2–6.5 dedicated to applied research and engineering design leading to an operational system. The HD program is strictly an applied research program, focused on development (corresponding to the 6.2 and 6.3 R&D levels) of concepts for which the critical basic research questions already have been answered. The existing funding allocation makes sense given the program's applied nature. However, the knowledge necessary for engineering development of a multimodal sensor of the type recommended in this report does not exist and would need to be developed through basic research.

The Army Research Office funded a five-year basic research effort through the Department of Defense Multidisciplinary University Research Initiative (MURI) from 1996 to 2001. It provided a total of $3.2 million per year to three university consortia for basic research on a variety of mine detection technologies. Some of this funding was for development of theory to support fusion of information from multiple sensors. However, a comprehensive, predictive theory for fusing signals or information from multiple landmine sensors still is not developed (J. Harvey, personal communication, 2002). Like other MURIs, this initiative covered only a five-year period and is not being continued. Current Army Research Office MURI investments, initiated in 2002, are funding three consortia a total of $9.2 million over the next five years (about $1.8 million per year) to explore three areas related to mine detection: (1) real-time explosive-specific chemical sensors, (2) the science of land target spectral signatures, and (3) detection and classification algorithms for multimodal inverse problems (UXO Center of Excellence, 2002). While some of the results of these efforts may be applied to mine detection, mine detection is not the exclusive focus of these new MURI projects.

The majority of U.S. funding for mine detection research is allocated to development of systems for countermine warfare. In 2002, the federal government invested $106 million in countermine research, about $75 million of which was spent on detection technologies (UXO Center of Excellence, 2002). The humanitarian demining community could leverage the technology developed for countermine operations, although some of the requirements are significantly different. (For example, detectors for countermine operations need not achieve the near-perfect probabilities of detection required for humanitarian demining.) The distribution of funds for countermine research resembles that for humanitarian mine detection in that the funding is concentrated on traditional GPR and EMI systems because of the emphasis on applied research.

As in the humanitarian demining program, the knowledge needed to field a multisensor system that yields one signal has not been developed through the countermine program. As discussed earlier, the development of HSTAMIDS, which combines GPR and EMI, has been a major focus of countermine research. However, because HSTAMIDS was developed in response to the immediate need for a system that works better than traditional metal detectors, it does not optimize use of the signals from the individual sensors. It combines separately designed EMI and GPR technologies on a single platform but does not necessarily represent the optimal combination of these two systems. Further, the output it produces is not aided by multisensor decision algorithms. The operator hears two separate signals from the EMI and GPR sensors and is not assisted in deciding which combination is most likely to indicate the presence of a mine. The decision about whether to declare an item a mine is not straightforward.

In some cases, strong signals from both the EMI and GPR detectors signal a mine, but in other cases the absence of a signal from one in conjunction with a strong signal from the other would indicate that a mine may be present. For example, consider a low-metal mine buried in dry sand with wet spots in the sand. An EMI detector will sense the currents induced in the metal or in the wet spots as being "targets." A GPR detector will identify a strong return signal from the wet sand and a relatively weak return signal from the plastic mine case, which has a similar dielectric constant as dry sand. In this case, the presence of a strong EMI response and the absence of a GPR response would indicate the presence of a target. This example illustrates the need for sensor fusion algorithms to assist the deminer in making declaration decisions. HSTAMIDS does not include such algorithms.

In sum, currently there is no basic research program focused on developing the fundamental knowledge needed to engineer novel multisensor detection systems for humanitarian demining. The funding allocated for humanitarian mine detection R&D is limited, and it is not dedicated to research toward a multisensor system. R&D conducted under countermine programs is well funded but is not optimized for multisensor system development. No basic research program exists to continue development of the theory needed to support a multisensor system.

## RECOMMENDED PROGRAM FOR PRODUCING AN ADVANCED MULTISENSOR SYSTEM

Development of an advanced multisensor mine detection system will require new, targeted R&D funds. The program should focus on the following areas:

1. algorithms for fusion of data from the individual sensors (to develop the theory necessary to support an advanced multisensor system);
2. integration of component systems (to address engineering issues associated with combining multiple sensors as part of a single device);
3. detection of the chemical components of explosives (to further develop components of the multisensor system that would search for explosives rather than for the mine casing and mechanical components); and
4. understanding of the fundamental physics of how the soil conditions in the shallow subsurface environment affect different sensors (to allow the development of models to predict performance across a range of environments).

This should be a long-term program, continued until the basic scientific and engineering knowledge needed to field an integrated, multisensor system is complete.

Because the long-term program will require a sustained commitment of substantial resources, we recommend that a short, relatively inexpensive preliminary study be conducted. The study should not focus on new research; rather, it should consolidate the existing empirical and theoretical work on sensor fusion and signal processing, with a focus on applications in landmine detection. The results of such a comprehensive literature review and assessment will help to narrow the focus of the long-term program we recommend to the most-promising avenues. This report covers some of the most-promising results to date; however, signal processing and sensor fusion are rapidly growing research areas and are being pursued by a broad array of academic, industrial, and governmental organizations in a diversity of disciplines. The resulting fragmentation of key empirical results and theoretical investigations makes it difficult to distinguish the rhetoric of sensor fusion and signal processing from the reality. Prior to investing in the long-term program, we believe it is crucial for decisionmakers to have a comprehensive assessment of the potential benefits and limitations of signal processing and sensor fusion for landmine detection, expert-informed guidance about which combinations of technologies and algorithms are most promising, and realistic estimates of the potential for improved operating characteristics.

## COST OF DEVELOPING A MULTISENSOR SYSTEM

As in any R&D initiative, the costs of developing a multisensor system are difficult to predict in advance. If the path forward for developing a next-generation detector were clearly defined, this would no longer be a research issue but strictly an engineering problem. Estimates of total anticipated research costs can be made based on prior experience, but it is important to keep in mind that these estimates may change over time as additional knowledge is gained.

To predict what investment might be needed to develop the next-generation mine detector, we used actual R&D costs of HSTAMIDS as a model. Table 3.3 shows actual costs of the various stages of HSTAMIDS development in the third column. The fifth column shows predicted costs of developing the next-generation multisensor system (excluding the costs of the preliminary study to consolidate existing research). The predictions for the next-generation detector are based on two assumptions:

1. that basic research to develop the new system will require three times as many researchers as participated in HSTAMIDS development, and

**Table 3.3**

**Estimated Costs and Time Required to Develop a Next-Generation Multisensor Landmine Detector**

| R&D Stage | Time Required for HSTAMIDS | Cost for HSTAMIDS | Estimated Time for Next-Generation Multisensor System | Estimated Cost for Next-Generation Multisensor System |
|---|---|---|---|---|
| Basic Research | 4 years (1990–1994) | $5 million | 5–8 years | $50 million |
| Prototype Development | 2 years (1994–1996) | $8 million | 2 years | $10 million |
| Demonstration and Validation | 5 years (1996–2001) | $33 million | 5 years | $40 million |
| Engineering and Manufacturing Development | 4 years (2001–2005) | $27 million | 4 years | $35 million |

2. that once the basic research is completed, the remaining stages of engineering the new system should cost approximately the same as the equivalent stages of building HSTAMIDS, adjusted by 1.1 percent per year for inflation.

The basic research for the next-generation detector will be considerably more complicated than that required for HSTAMIDS for two primary reasons. First, HSTAMIDS incorporates two mature technologies (EMI and GPR), whereas the next-generation system might include newer detection methods, such as NQR and/or acoustic/seismic sensors. Performance of these newer methods cannot be optimized or predicted without additional research. Second, HSTAMIDS does not include advanced algorithms for sensor fusion. The theoretical basis needed to accomplish advanced sensor fusion is not yet complete. As Table 3.3 shows, we expect that eight years of basic research will be required to support an advanced, multisensor detector, but it is possible that this research could be compressed into five years, with the total amount of funding remaining the same but with a higher amount spent each year. A prototype system could be available within two years of completing the basic research.

Table 3.4 summarizes the types of basic research required to support development of a next-generation, multisensor mine detector, as

**Table 3.4**

**Itemization of Basic Research Requirements and Costs over the Next Five Years for a Next-Generation Multisensor Mine Detector**

| Research Area | Estimated Number of Researcher Years over the Next Five Years | Estimated Cost over the Next Five Years | Anticipated Results After Five Years of Research |
|---|---|---|---|
| Algorithms for sensor fusion | 40 | $10.00 million | Minimal set of sensor-level fusion algorithms for specific sensor suite |
| Integration of component sensor technologies | 25 | $6.25 million | Multisensor prototype detector with three to four sensor technologies |
| Explosives detection technologies | 50 | $12.50 million | Set of sensors suitable for integration in multisensor prototype; three to five would be candidates for immediate integration, with remainder used for backups or requiring long-term research for midlife upgrades |
| Subsurface environment effects on sensors | 10 | $2.50 million | Understanding of major soil parameters that affect the sensors identified above; set of simple tests that can be performed in situ to provide information to improve or predict performance of those sensors |

recommended in this report. Research costs are based on the judgment and experience of the S&TPI/OSTP Landmine Detection Technology Task Force. The amounts shown here assume that the basic research work is not compressed—i.e., that significant progress will occur after five years, but that funding is not sufficient to complete the work in this time frame.

As shown in Tables 3.3 and 3.4, the total estimated costs for basic research toward developing a next-generation detector are $50 million over five to eight years, or $6.25–10.00 million per year. (The higher annual figure assumes that the research would be compressed into five years.) The total anticipated costs, including basic research, prototype development, demonstration and validation, and engineering and manufacturing development are $135 million over 16–19 years ($7.1–8.4 million per year). As shown, a prototype system could be available within seven years at a total cost of $60 million ($8.6 million per year) if the basic research were compressed with higher up-front spending.

## SUMMARY OF RECOMMENDATIONS

In sum, no basic research program focused on developing the fundamental knowledge needed to engineer novel multisensor detection systems for humanitarian demining currently exists. We recommend that the United States invest in developing such a system to enable more rapid, safer clearance of antipersonnel mines. Such a system would save billions of dollars in the cost of mine clearance, and the resulting advances in signal processing and sensor fusion would be transferable to many other disciplines and applications.

The first phase of the multisensor mine detection program should be a limited, proof-of-concept study (costing less than $1 million) that would consolidate existing research on sensor fusion and signal processing. This study would identify the most-promising directions for multisensor development and the key information gaps.

Once the proof-of-concept study is completed, the full multisensor development initiative would begin with research focused on the following broad areas:

- algorithmic fusion of data from individual sensors ($2.0–3.2 million per year);
- integration of component sensor technologies ($1.25–2.00 million per year);
- detection of chemical components of explosives ($2.5–4.0 million per year); and
- modeling of the effects of soil conditions in the shallow subsurface on sensor technologies ($0.5–0.8 million per year).

# REFERENCES

Ackenhusen, J. G., Q. A. Holmes, et al., *Detection of Mines and Minefields*, Ann Arbor, Mich.: Veridian Systems Division, 2001.

Andersson, N., C. P. D. Sousa, et al., "Social Cost of Land Mines in Four Countries: Afghanistan, Bosnia, Cambodia, and Mozambique," *British Medical Journal*, No. 311, September 16, 1995, pp. 718–721.

Blagden, P. M., "Kuwait: Mine Clearing After Iraqi Invasion," *Army Quarterly & Defence Journal*, Vol. 126, No. 1, 1996, pp. 4–12.

Cameron, M. A., R. J. Lawson, et al., *To Walk Without Fear: The Global Movement to Ban Landmines*, Toronto: Oxford University Press, 1998.

Carruthers, A., and J. McFee, "Landmine Detection: An Old Problem Requiring New Solutions," *Canadian Defence Quarterly*, Vol. 25, No. 4, 1996, pp. 16–18.

Collins, L. M. "EMI, GPR and MAG Signal Processing and Fusion," presented at Research on Demining Technologies Joint Workshop, Ispra, Italy, July 12–14, 2000.

Collins, L. M., L. G. Huettel, et al., "Sensor Fusion of EMI and GPR Data for Improved Land Mine Detection," in *Detection and Remediation Technologies for Mines and Minelike Targets VII*, J. T. Broach, R. S. Harmon, and G. J. Dobeck, eds., Seattle: International Society for Optical Engineering, 2002.

Cumming, C., C. Aker, et al., "Using Novel Fluorescent Polymers as Sensory Materials for Above-Ground Sensing of Chemical Signature Compounds Emanating from Buried Landmines," *IEEE Transactions on Geoscience and Remote Sensing*, Vol. 39, No. 6, 2001, pp. 1119–1128.

Das, Y., J. T. Dean, et al., *International Pilot Project for Technology Co-operation Final Report: A Multinational Technical Evaluation of Performance of Commercial Off the Shelf Metal Detectors in the Context of Humanitarian Demining*, Ispra, Italy: European Commission Joint Research Center, 2001.

Dasarathy, B. V., *Decision Fusion*, Los Alamitos, Calif.: IEEE Computer Society, 1994.

Doleman, B. J., and N. S. Lewis, "Comparison of Odor Detection Thresholds and Odor Discriminabilities of a Conducting Polymer Composite Electronic Nose Versus Mammalian Olfaction," *Sensors and Actuators*, No. 72, 2001, pp. 41–50.

Freund, M. S., and N. S. Lewis, "A Chemically Diverse Conducting Polymer-Based 'Electronic Nose,'" *Proceedings of the National Academy of Sciences*, No. 92, March 1995, pp. 2652–2656.

Gunatilaka, A. H., and B. A. Baertlein, "Feature-Level and Decision-Level Fusion of Noncoincidently Sampled Sensors for Land Mine Detection," *IEEE Transactions on Pattern Analysis and Machine Intelligence*, Vol. 23, No. 6, 2001, pp. 577–589.

Hewish, M., and R. Pengelley, "Treading a Fine Line: Mine Detection and Clearance," *Jane's International Defense Review*, Vol. 30, No. 11, 1997.

Hopkins, A. R., and N. S. Lewis, "Detection and Classification Characteristics of Arrays of Carbon Black/Organic Polymer Composite Chemiresistive Vapor Detectors for the Nerve Agent Simulants Dimethylmethyphosphonate and Diisopropylmethylphosphonate," *Analytical Chemistry*, Vol. 73, No. 5, 2001, pp. 884–892.

Horowitz, P., K. Case, et al., *New Technological Approaches to Humanitarian Demining*, McLean, Va.: MITRE, JSR-92-115, 1996.

Hubert, D., "The Challenge of Humanitarian Mine Clearance," in *To Walk Without Fear: The Global Movement to Ban Landmines*, M. Cameron, R. J. Lawson, and B. W. Tomlin, eds., Toronto: Oxford University Press, 1998, pp. 315–335.

International Campaign to Ban Landmines (ICBL), *Landmine Monitor Report 2001*, Washington, D.C., 2001.

Jeffrey, S. J., "Antipersonnel Mines: Who Are the Victims?" *Journal of Accident and Emergency Medicine*, Vol. 13, No. 5, 1996, pp. 343–346.

Johnston, J. M., M. Williams, et al., "Canine Detection Odor Signatures for Mine-Related Explosives," in *Detection and Remediation Technologies for Mines and Minelike Targets III*, A. C. Dubey, J. F. Harvey, and J. Broach, eds., Seattle: International Society for Optical Engineering, 1998.

Kercel, S. W., R. S. Burlage, D. R. Patek, C. M. Smith, A. D. Hibbs, and T. J. Rayner, "Novel Methods of Detecting Buried Explosive Devices," in *Detection and Remediation Technologies for Mines and Minelike Targets II*, A. C. Dubey and R. L. Barnard, eds., Seattle: International Society for Optical Engineering, 1997.

Koh, G., "Effect of Soil Moisture on Radar Detection of Simulant Mines," fact sheet, Hanover, N.H.: U.S. Army Corps of Engineers, Cold Regions Research and Engineering Laboratory, 1998.

Lewis, A. M., P. S. Verlinde, et al., "Recent Progress in the Joint Multisensor Mine-Signatures Database Project," in *Detection and Remediation Technologies for Mines and Minelike Targets VII*, J. T. Broach, R. S. Harmon, and G. J. Dobeck, eds., Seattle: International Society for Optical Engineering, 2002.

Office of the Assistant Secretary of Defense for Special Operations/Low Intensity Conflict, *Landmine Casualty Data Report: Deminer Injuries*, Washington, D.C.: U.S. Department of Defense, 2000.

Olhoeft, G. "Ground Penetrating Radar: Introduction and History," available at www.g-p-r.com/introduc.htm (last accessed October 15, 2002).

Phelan, J. M., and J. L. Barnett, "Chemical Sensing Thresholds for Mine Detection Dogs," in *Detection and Remediation Technologies for Mines and Minelike Targets VII,* J. T. Broach, R. S. Harmon, and G. J. Dobeck, eds., Seattle: International Society for Optical Engineering, 2002.

Rappaport, C., S. Winton, D. Jin, and L. Siegal, "Modeling the Effects of Non-Uniform Soil Moisture on Detection Efficacy of Mine-Like Objects with GPR," in *Detection and Remediation Technologies for Mines and Minelike Targets IV,* A. C. Dubey, J. F. Harvey, J. Broach, and R. E. Dugan, eds., Seattle: International Society for Optical Engineering, 1999.

Rosen, E. M., K. D. Sherbondy, et al., "Performance Assessment of a Blind Test Using the University of Mississippi's Acoustic/Seismic Laser Doppler Vibrometer (LDV) Mine Detection Apparatus at Fort A. P. Hill," in *Detection and Remediation Technologies for Mines and Minelike Targets V,* A. C. Dubey, J. F. Harvey, J. Broach, and R. E. Dugan, eds., Seattle: International Society for Optical Engineering, 2000.

Rotondo, F., T. Altshuler, et al., *Report on the Advanced Technology Demonstration (ATD) of the Vehicular-Mounted Mine Detection (VMMD) Systems at Aberdeen, Maryland, and Socorro, New Mexico,* Alexandria, Va.: Institute for Defense Analyses, 1998.

Treveylan, J., "Practical Issues in Manual Demining: Implications for New Detection Technologies," in *Detection of Explosives and Landmines: Methods and Field Experience,* H. Schubert and A. Kuznetsov, Dordrecht, The Netherlands: Kluwer Academic Publishers, 2002a, pp. 155–164.

Treveylan, J., "Technology and the Landmine Problem: Practical Aspects of Mine Clearance Operations," in *Detection of Explosives and Landmines: Methods and Field Experience,* H. Schubert and A. Kuznetsov, Dordrecht, The Netherlands: Kluwer Academic Publishers, 2002b.

U.S. Department of State, *To Walk the Earth in Safety: The United States Commitment to Humanitarian Demining,* Washington, D.C.: U.S. Department of State, Bureau of Political-Military Affairs, 2001.

U.S. General Accounting Office (GAO), *Mine Detection: Army Detector's Ability to Find Low-Metal Mines Not Clearly Demonstrated*, GAO/NSIAD-96-198, 1996.

UXO Center of Excellence, *UXO Center of Excellence Annual Report for 2001*, Ft. Belvoir, Va.: Joint Unexploded Ordnance Coordination Office, 2002.

Vines, A., and H. Thompson, *Beyond the Landmine Ban: Eradicating a Lethal Legacy*, London: Research Institute for the Study of Conflict and Terrorism, 1999.

Witten, T. R., P. R. Lacko, et al., "Fusion of Ground Penetrating Radar and Acoustics Data," in *Detection and Remediation Technologies for Mines and Minelike Targets VII*, J. T. Broach, R. S. Harmon, and G. J. Dobeck, eds., Seattle: International Society for Optical Engineering, 2002.

Yang, J. S., and T. M. Swager, "Porous Shape Persistent Fluorescent Polymer Films: An Approach to TNT Sensory Materials," *Journal of the American Chemical Society*, No. 120, 1998, pp. 5321–5322.

# AUTHOR AND TASK FORCE BIOGRAPHIES

## AUTHORS

**Jacqueline MacDonald** is an engineer at RAND. She specializes in analysis of environmental remediation issues. Prior to joining RAND in 1999, Ms. MacDonald served as associate director of the National Research Council's Water Science and Technology Board. She worked there for nearly a decade directing and managing studies related to cleanup of contaminated groundwater and soil, remediation of sites in the former nuclear weapons complex, and management of municipal water supplies. She has published extensively in environmental journals, including *Environmental Science & Technology, Water Research, Journal of the American Water Works Association,* and *Water Environment & Technology*. Ms. MacDonald earned her M.S. in environmental science in civil engineering from the University of Illinois at Urbana-Champaign and her B.A. magna cum laude in mathematics from Bryn Mawr College.

**J. R. Lockwood** has been an associate statistician in the RAND Statistics Group since 2001. He specializes in quantitative analyses of environmental and education policy problems. His methodological interests include Bayesian model complexity and model selection, hierarchical and spatial modeling, and Markov chain Monte Carlo methods. He holds an M.S. and Ph.D. in statistics from Carnegie Mellon University and a B.A. summa cum laude in environmental science and policy from Duke University. He received the 2001 Leonard J. Savage Award for the outstanding doctoral dissertation in Bayesian Application Methodology.

## TASK FORCE CHAIR

**John McFee** has been leading and conducting research in the detection of mines, minefields, and UXO for 25 years. At present, he is responsible for developing, directing, and executing the mine detection research programs for the Canadian military and the Canadian Centre for Mine Action Technology. Researchers in other countries have adopted a number of his research group's detection algorithms, models, concepts, and configurations. Among his group's achievements are development of the first multisensor, teleoperated, and vehicle-mounted mine detector fielded by a military and the first real-time hyperspectral detection of mines. His group has also worked in the field in Cambodia, Bosnia, Croatia, Afghanistan, and Colombia to solve practical problems with in-service mine detectors and to determine which detectors were best suited for each particular location. Dr. McFee's research has involved a wide range of technologies to detect mines, minefields, and UXO, including magnetometers, EMI systems, pattern classification, image analysis, and passive and active infrared imaging. He presently concentrates his R&D on hyperspectral imaging and nuclear detection of mines, including TNA, neutron albedo imaging, and x-ray imaging. Since 1995, he either has cochaired or has been a program committee member of the annual International Society for Optical Engineering (SPIE) Conference on Detection and Remediation of Mines and Minelike Targets. Dr. McFee received the B.Sc. honours physics degree from the University of New Brunswick and a Ph.D. degree in nuclear physics from McMaster University.

## TASK FORCE MEMBERS

**Thomas Altshuler** is a research staff member at IDA. From 1998 to 2002 he took a leave from IDA to serve as a program manager at DARPA, where he managed several programs, including the DARPA Dog's Nose/UXO Detection Program during its final phase as well as the Antipersonnel Landmine Alternative Program. His work at IDA has included supporting mine and UXO detection programs for DARPA and other Department of Defense programs. This work has included technology assessment and evaluation of detection systems under development, as well as oversight of experiments investigating environmental and man-made clutter on fielded mine and UXO detection technology. Dr. Altshuler holds a Ph.D. from the Massachusetts Institute of Technology, where his work focused on magnetic materials.

**J. Thomas Broach** began his R&D career in 1969 by investigating the application of superconductivity to specialized electric power sources. He began working in the development of mine detection technologies in 1983 when he joined the countermine division of the U.S. Army Night Vision and Electronic Sensors Directorate (NVESD). From 1986 to 1997, Dr. Broach was head of the Detection Research Team, which was responsible for developing various mine detection technologies for use in handheld, vehicular-mounted, and remote detection applications. During this time, he directed mine detection research, which included such sensors as GPR, acoustic/seismic, nuclear, x-ray, and electro-optic. In 1996, he was appointed by the Army Countermine Task Force to field test and evaluate 13 detectors for their potential for immediate troop use. He was in charge of the U.S. Army Vehicular Mounted Mine Detector Advanced Technology Demonstration, which was completed in 1998. Since 1998, he has chaired the annual SPIE Conference on Detection and Remediation Technologies for Mines and Minelike Targets. Currently, Dr. Broach is senior scientist of the countermine division at NVESD. He holds a Ph.D. in physics from American University.

**Lawrence Carin** is a professor of electrical engineering at Duke University. Dr. Carin was a lead investigator for the recently completed MURI on the detection of landmines. His research has focused on the basic physics and chemistry associated with a wide range of landmine sensors. He earned his Ph.D. in electrical engineering from the University of Maryland.

**Russell Harmon** is branch chief for terrestrial sciences at the Army Research Office (ARO) and currently manages the ARO's basic research program in landmine detection. He is a geochemist who has worked in the Lunar Receiving Laboratory and Geochemistry Division at the NASA Manned Spacecraft Center, the Scottish Universities Research and Reactor Center, and the UK Natural Environment Research Council. He has held faculty positions at Michigan State University and Southern Methodist University. Dr. Harmon is a current chair of the annual SPIE Conference on Detection and Remediation of Mines and Minelike Targets. He holds a B.A. from the University of Texas, an M.S. from Pennsylvania State University, and a Ph.D. from McMaster University.

**Carey Rappaport** is a professor of electrical and computer engineering at Northeastern University. Dr. Rappaport was a principal investigator for the recently completed MURI on landmine detection. He is currently coprincipal investigator and associate director of CenSSIS, the Center for Subsurface Sensing and Imaging Systems, a National Science Foundation engineering research center. His research areas include antenna design, finite difference numerical analysis of scattering, and analysis of wave propagation and scattering in complex dielectric media. He earned his S.B. in mathematics and his S.B., S.M., E.E., and Ph.D. degrees in electrical engineering from the Massachusetts Institute of Technology.

**Waymond Scott** is a professor of electrical engineering at Georgia Tech. His research involves the interaction of electromagnetic and elastic waves with materials. This research spans a broad range of topics, including the measurement of the properties of materials, experimental and numerical modeling, and systems for the detection of buried objects. Currently, his research is concentrated on investigating techniques for detecting objects buried in the earth. This work has many practical applications and includes, for example, the detection of underground utilities, buried hazardous waste, UXO, and buried landmines. Dr. Scott received his B.S., M.S., and Ph.D. degrees in electrical engineering from Georgia Tech.

**Richard Weaver** has held a variety of positions in the U.S. Army's research community. He currently is the director of the countermine division at NVESD. He served as the first director of JUXOCO, which was established to coordinate all Department of Defense work concerning detection of UXO and landmines. He has published articles covering countermine research in *Geoscience and Remote Sensing* and in journals of the Institute of Electrical and Electronics Engineers. He has spoken about mine detection at numerous national and international symposia. He received a B.A. in chemistry from Bucknell University and an M.S. in systems engineering from the University of Southern California. Mr. Weaver is a veteran of the Vietnam War.

Appendix A

# ELECTROMAGNETIC INDUCTION (PAPER I)

*Yoga Das, Defence R&D Canada–Suffield*[1]

## OPERATING PRINCIPLE

Electromagnetic induction (EMI) is the basis for the familiar hand-held metal/mine detectors. The metal parts present in a landmine are detected by sensing the secondary magnetic field produced by eddy currents induced in the metal by a time-varying primary magnetic field. The frequency range employed is usually limited to a few tens of kHz. The primary field is produced by an electrical current flowing in a coil of wire (transmit coil), and the secondary field is usually detected by sensing the voltage induced in the same or another coil of wire (receive coil). Present research is investigating replacement of the receive coil with magnetoresistive devices. EMI detectors are often classified into two broad categories: "continuous wave" and "pulse induction."

This appendix discusses EMI detectors primarily in the context of minimum-metal antipersonnel landmines encountered in humanitarian demining. Words such as "EMI sensors," "EMI detectors," "metal detectors," and "metal/mine detectors" are used interchangeably throughout.

---

[1]Originally published by Defence R&D Canada–Suffield, 2001. Reprinted with permission.

## STATE OF DEVELOPMENT

The basic technology is very mature. The EMI principle appears to have been used for landmine detection in World War I. EMI metal/mine detectors were further developed during World War II and have been routinely used to detect landmines since. The use of EMI to detect conducting objects is also well established in other application areas such as mineral exploration, nondestructive testing, treasure hunting, security and law enforcement, and food processing.

Although earlier development of EMI mine detectors was led by government agencies (examples include the development of the 4C and the AN/PSS-11), recent development of this technology has been driven, in large part, by the commercial interest of the private sector. As a result, a number of detector models from various companies worldwide are available as commercial-off-the-shelf (COTS) items. However, most of these items were not developed to meet any specific statement of requirement for mine detection.

The private sector has carried out product improvement through its own research and development (R&D). Government organizations in various countries have continued to conduct or sponsor research in areas of EMI mine/unexploded ordnance detection, which are considered to have low probability of success but high potential payoff.

## CURRENT CAPABILITIES

Largely due to the advancement in electronics, the sensitivity, sophistication, and ergonomics of metal detectors have improved tremendously since their beginning in World War II. A good modern metal detector can detect extremely small quantities of metal (e.g., that found in an M14 or 72A antipersonnel landmine buried up to 10 cm) under various soil (e.g., magnetic) and other environmental (e.g., wet tropical) conditions. The speed at which a handheld metal detector can be used to sweep the ground is typically less than 1 m/second. However, the effective rate of area coverage will depend on many factors, which include the search halo of the detector, the frequency of occurrence of metal fragments and actual landmines, and the operating procedure employed.

The capabilities of COTS metal detectors vary over a wide range. Because most of these were not developed to meet any specific mine detection requirement and quality control standard, there is wide variability in the performance and quality among the various detector models. Developing and standardizing systematic testing and evaluation procedures to help select detectors best suited for a given set of circumstances will, in itself, represent a worthwhile contribution to mine detection.

## LIMITATIONS

The most obvious and serious limitation of metal detectors used to detect landmines is the fact that they are *metal* detectors. A modern metal detector is very sensitive and can detect tiny metal fragments as small as a couple of millimeters in length and less than a gram in weight. An area to be demined is usually littered with a large number of such metal fragments and other metallic debris of various sizes. This results in a high rate of "nuisance" alarms since a metal detector cannot currently distinguish between the metal in a landmine and that in a harmless fragment. The more sensitive a detector is, the higher the number of nuisance alarms it is likely to produce in a given location. Operating a detector at a lower sensitivity to reduce the number of such nuisance alarms may render it useless for detecting the very targets it was designed to detect, that is, the minimum-metal-content landmines buried up to a few centimeters.

Electromagnetic properties of certain soils can limit the performance of metal detectors. Only a few detectors in the market are designed to and can cope with magnetic soils without serious loss of sensitivity. But not all of these detectors can cope with magnetic soils to the same extent. Because magnetic soils are found in most parts of the world that have landmine problems, the inability to operate satisfactorily in such soils is a significant limitation of some metal/mine detectors. Further, performance of some detectors could be severely limited by certain other environmental factors, such as accumulation of moisture on the detector head. However, this is not a limitation of the basic EMI technology.

## POTENTIAL FOR IMPROVEMENT

Because metal detectors have been around for a long time and have been extensively researched, startling breakthroughs are not expected. It is doubtful if an order of magnitude improvement in metal detector performance is possible in the next two to seven years. However, sustained effort on a few fronts will contribute to a more efficient utilization of this technology in demining and to scientific progress in this area of sensing. A brief analysis of potential research areas, some of which are interrelated, and the expected results are described in the R&D outline that follows.

## RESEARCH AND DEVELOPMENT PROGRAM OUTLINE

This outline introduces the various potential research areas, makes a hypothesis about the possible improvement from each, recommends specific research goals, and indicates expected results. Because the author is not familiar with the cost of doing R&D in the United States, only a very crude guess of the level of personnel effort needed is included. As well, a level of priority on a scale of 1 to 3, with 1 being the highest, has been assigned to each research area with a brief explanation. This prioritization should help when a resource allocation choice has to be made.

### Nuisance Alarm Reduction

A major improvement in the utility of a metal detector would be achieved if the number of nuisance alarms could be reduced without sacrificing detectability of the landmines. The following three general approaches address this issue: signal processing, imaging, and multiple sensors. Of these, the use of multiple sensors would likely have the *most* immediate impact. However the topic of multiple sensing is beyond the scope of this report and hence an R&D plan for it will not be included here.

**Signal Processing.** The signal-processing approach consists of analyzing the EMI response waveform of an object to distinguish it from other objects. It is highly unlikely that this approach will ever result in a standalone metal detector capable of reliably differentiating between landmines and metal fragments in a field situation. However, advancements in this area, expected to be slow, may provide synergistic improvement to other approaches. Research in this area has lacked experimental emphasis—many techniques have been proposed, but very little experimental validation exists. Potential improvement offered by this approach can only be established through a program of extensive experimentation that would include measurement and analysis of EMI response of a large number of objects under various conditions.

*Recommended R&D:* Conduct an extensive experimental study to measure EMI response of a large number of targets of interest—first with the targets in air and then buried. The focus of this study would be to validate or establish the limitations of proposed discrimination techniques. This study will also establish whether current EMI sensor technology is capable of providing the quality of data needed to apply these techniques.

*Expected results:* Antipersonnel landmine/metal fragments discrimination is considered high-risk and long-term research, and as such it is not expected that a fieldable method would emerge in the two-to-seven year time frame. However, the recommended research will answer important questions on the viability of the signal-processing approach.

*Recommended level of effort:* three to four person years (PYs) for five to seven years. Personnel: technologists, engineers, and scientists.

*Priority:* 2. This area has been researched for a while and past progress has been slow. However, this research should be sustained.

## Imaging

Producing images, even crude ones, of the detected objects would reduce nuisance alarms. Even the ability to image some of the larger pieces near the surface would help. Imaging with EMI sensors is in its infancy. Some metal targets have been imaged in the laboratory using data from a metal detector scanned on a plane over the object. EMI imaging techniques, using data from an array of sensors, have also been proposed to image faults in nondestructive testing of metal parts. A basic requirement for imaging is the availability of response measurements as a function of sensor position. This can be achieved either by knowing the position of a single sensor as it is moved or by using a suitable array of sensors. Magnetoresistive sensor arrays appear to be an attractive option for the latter approach.

*Recommended R&D:* Investigate both of the above approaches to imaging for their applicability to landmine detection. This will entail both theoretical and experimental studies. Data gathering with a single detector as well as with proposed magnetoresistive sensor arrays should be investigated.

*Expected results:* As with discrimination through signal processing, the imaging approach is also high risk and long term. A fieldable imaging system is not expected in the two-to-seven-year time frame. However, this research should answer some crucial questions regarding this approach, such as: Are current sensors, including position sensors, capable of providing data needed to produce an image? What are the limits on the size and depth of objects for which such data can be obtained? How quickly can a needed data set be gathered and an image produced?

*Recommended level of effort:* three to four PYs for five to seven years. Personnel: technologists, engineers, and scientists.

*Priority:* 1. This area has not been explored much. If successful, the potential benefits would be high.

## Enabling Technologies

**New Sensors.** Magnetoresistive sensors and sensor arrays may provide an attractive alternative to the conventional EMI sensors and merit a separate investigation. Potential advantages of this technology include small size, high bandwidth, and availability of an array with a large number of elements. Both individual and arrays of magnetoresistive sensors have been investigated to replace conventional wire coils. Work on sensor arrays has stopped over the past couple of years.

*Recommended R&D:* (a) Further investigate the development and testing of magnetoresistive sensor arrays[2] for landmine detection,

---

[2]Previously investigated by Blackhawk Geometrics and NVE.

and (b) characterize the performance limits of a single sensor as a replacement for a wire coil in a metal detector.

*Expected results:* This research will establish whether current magnetoresistive sensors and sensor arrays can provide signals of sufficient strengths from targets of interest in antipersonnel landmine detection. If they can, such sensors will be of great help to both signal-processing and imaging efforts already described.

*Recommended level of effort:* one to two PYs for two years. Personnel: technologists, engineers, and scientists. Development of the basic technology of magnetoresistive sensors will continue to benefit from a number of other applications.

*Priority:* 1. This area has not been sufficiently explored. This is truly a new area of research in EMI detection. In spite of known problems, this area should be explored extensively because the potential benefits would be high.

**Positioning Systems.** Light, inexpensive, and compact positioning systems that can be integrated to a handheld detector will enable those detectors to provide real-time EMI data as a function of sensor position. Such data are needed for imaging and target localization. A few systems have been proposed to address this requirement.

*Recommended R&D:* Conduct a critical review of proposed systems and develop, if needed, light, compact, and inexpensive positioning systems to integrate with metal detectors.

*Expected results:* This research is considered low risk and short term. A suitable positioning system will greatly help target localization and imaging efforts.

*Recommended level of effort:* one to two PYs for two years. Personnel: technologists and engineers.

*Priority:* 2. There is some ongoing research in other countries that one may be able to harness.

## Speed of Coverage Improvement

Current practice involves an individual deminer sweeping with a single detector and localizing targets manually using hand, ear, and eye coordination. Use of wide detector arrays as well as quick and accurate target localization should speed up the overall demining process.

*Recommended R&D:* (a) Develop wide detector arrays further to make them suitable for antipersonnel landmine detection, consider the possibility of person-portable arrays, and improve target localization algorithms for arrays; and (b) develop and integrate automatic target localization into current handheld systems possibly including overlaying of target positions with a photo of the scanned area. This research should be coordinated with research on positioning systems.

*Expected results:* This should be considered low risk and medium term. With proper focus, this research should result in fieldable improvements.

*Recommended level of effort:* On (a) three to four PYs for four to five years. On (b) one to two PYs for two years (this part would gain leverage from effort on positioning systems). Personnel: technologists, engineers, and scientists.

*Priority:* 2. Other countries are looking at some aspects of this.

## Operational Improvement

Capabilities of available detectors vary widely. Proper equipment selection will make sure that the best that current EMI technology can offer is used in demining. To this end, development of an internationally accepted standard for scientific testing of EMI detectors for landmines will help improve quality and possibly speed of demining. The developed standard must include a method of unambiguously defining a "detection" that accounts for the effect of the operator. Normally an operator makes a "detection" or "no detection" decision by listening to the audio output from a detector. Such a process is obviously prone to variability introduced by the operator.

*Recommended R&D:* (a) Critically review existing test methodologies proposed over the years and contribute to current international efforts to coordinate development of a standard for metal detector testing for humanitarian demining. (This is considered low-risk but high-priority work.) (b) As a part of (a) or as an independent effort, develop computer processing techniques and/or experimental protocols to account for the influence of the operator.

*Expected results:* An objective standard for scientific testing of metal detectors used in humanitarian demining will help improve the average quality of COTS detectors.

*Recommended level of effort:* two to four PYs for two years. Personnel: engineers and scientists.

*Priority:* 1. This "low-tech" effort will have the most immediate impact on product improvement by the private sector and on humanitarian demining.

## Soils Study

**Analytical and Experimental Study.** EMI sensors are affected by the electromagnetic properties of soil, particularly the magnetic susceptibility. Although a few detectors can cope with magnetic soils to a degree, the influence of soil electromagnetic properties on EMI sensors is not fully understood. Such an understanding will help us predict as well as improve the performance of EMI detectors in magnetic soils.

*Recommended R&D:* Conduct analytical and experimental study of the effect of soil electromagnetic properties (magnetic susceptibility and electrical conductivity) on EMI sensors in the specific context of landmine detection. An integral part of this work should be selection and/or development of suitable instrumentation to measure relevant soil properties.

*Expected results:* This research will involve significant intellectual effort from specialists and should produce: (1) a model to predict the sensor response as a function of soil electromagnetic properties and (2) a guide to characterizing soils in terms of their electromagnetic properties to help compare performance of different detectors in various soils. Results of this will also feed the effort on development of testing standards.

*Recommended level of effort:* three to five PYs for at least five years. Personnel: scientists and technologists.

*Priority:* 1. This long overdue and challenging research will provide important information of practical value.

**Soil Database.** An information database (and/or a soil map) of the electromagnetic properties of the top 50-cm layer[3] of soil of the landmine-affected regions of the world will be a very valuable resource. Such information will help planning of current demining operations as well as research on landmine detection.

*Recommended R&D:* Initiate and/or contribute to international efforts to develop an information database of the electromagnetic properties (initially magnetic susceptibility and electrical conductivity) of the top layer of soil in landmine-affected parts of the world.

*Expected results:* This will likely be an ongoing long-term effort. The information as it is gathered will populate an electronic database and will be immediately useful to researchers and planners.

*Recommended level of effort:* Cost will depend on scope of the project. Personnel: scientists, engineers, and technologists of multiple disciplines.

*Priority:* 3. This will be a project of large scope and some initial effort will go toward planning.

## FURTHER READING

A vast literature exists on the many aspects, including field testing, of EMI detectors. The following is the latest multinational report describing comprehensive testing of COTS metal detectors for use in humanitarian demining:

Y. Das et al., eds., *Final Report of the International Pilot Project on Technology Co-operation (IPPTC) for the Evaluation of Metal/Mine Detectors*, EUR 19719 EN, June 2001 (published on behalf of the participants by the European Commission, Joint Research Centre, Ispra, Italy).

[3]Such a database will also help R&D on other sensors.

Appendix B

# ELECTROMAGNETIC INDUCTION (PAPER II)

*Lloyd S. Riggs, Auburn University*

## INTRODUCTION

Electromagnetic induction (EMI)–based detection techniques find application in a variety of areas, including nondestructive testing (e.g., locating cracks in turbine blades), ore body location, as well as the detection and *identification* of landmines and unexploded ordnance (UXO). The common metal detector is probably the most ubiquitous EMI device in use today.

Typical components of a metal detector include a transmitter and receiver coil. As depicted in Figure B.1, electric currents that flow in the transmitter coil radiate a primary magnetic field that penetrates the surrounding medium and any nearby metallic object. A time-changing primary magnetic field will induce so-called eddy currents in the buried object, and these currents in turn radiate a secondary magnetic field that is sensed (picked up) by the receiver coil. An audio tone is produced whenever the metallic object causes the induced receiver coil voltage to exceed some threshold.

Modern metal detectors are quite sensitive and can detect buried low metallic (LM)–content landmines that contain only a few grams of metal [1]. Examples of modern-day metal detectors include the U.S. Army's standard-issue AN/PSS-12 manufactured by Schiebel Corporation of Austria and the F3 manufactured by Minelab Corporation of Australia. Unfortunately, as the name implies, a metal detector will produce an audio alarm whenever *any* metallic object is brought near its search coil(s). At present, commercially available metal

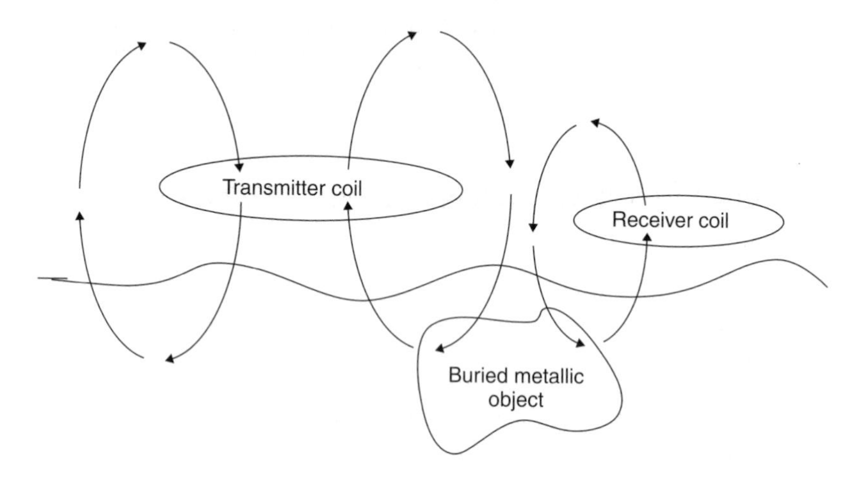


**Figure B.1—Typical Electromagnetic Induction System**

detectors have very limited ability to discriminate between landmines and buried metallic clutter. False alarms generated by metallic clutter severely limit the speed and efficiency of mine clearance operations. For example, Minelab, whose EMI instrumentation is frequently used in humanitarian demining operations, reports that it is not uncommon to remove 1,000 metallic clutter items per mine. The goal for modern EMI systems is, however, more than detection. In order to discriminate clutter from LM mines, additional information has to be gathered about the target.

The additional information that is available with an EMI sensor is contained in the details of the orthogonal mode structure of the eddy currents and associated induced fields, and how they evolve over time. The eddy current modes are related to the eigenvalues (referred to as "response coefficients," or "βs") of the magnetic polarizability tensor [2,3]. EMI sensors are generally characterized as pulse induction or continuous wave, with the former using, as the name implies, short pulses of current in the transmitter coil while the latter forces a continuous sinusoidal current to flow in the transmitter coil. Both are capable of measuring how the eddy currents evolve over time, yielding time or frequency-dependent βs that are related to each other by Fourier transforms, and each technique has advantages and disadvantages [4]. The GEM-3 manufactured by Geophex is an example of a continuous wave detector, and the AN/PSS-12 and F3 mentioned above are examples of pulse-induction detectors. Discrimination experiments with the AN/PSS-12 will be described in the next section.

## ACHIEVING DISCRIMINATION CAPABILITY WITH A MODIFIED AN/PSS-12 [5]

EMI discrimination research has focused on either the UXO or the mine problem. In the former case, one is generally interested in discriminating between large unexploded bombs and surface clutter while in the latter, as described above, the problem is one of discriminating between LM-content landmines and clutter. UXO can be buried several meters deep while antipersonnel mines are typically buried no more than 6 inches below the surface. EMI sensors designed to detect UXO usually incorporate rather large coils (on the order of a meter in diameter) while sensors designed to detect LM mines usually employ coils less than 12 inches in diameter. It should be mentioned that large LM "plastic" anti-vehicular mines are frequently more difficult to detect (and therefore discriminate from metallic clutter) than smaller LM antipersonnel mines. This is because they both contain approximately the same amount of metal (usually a small metallic detonation tube and firing pin), but the antivehicular mine is usually buried deeper than the antipersonnel mine. For explanatory purposes, this discussion will focus on the mine problem emphasizing modifications to the Army's AN/PSS-12 metal detector in order to enhance landmine discrimination capability.

Although the AN/PSS-12 is quite sensitive and can detect very small amounts of metal, it was not designed to discriminate among metallic objects, or, more specifically, between metallic clutter and LM "plastic" landmines. It has been shown that the response of a metallic object is characterized by distinct real-axis poles in complex frequency domain or equivalently by a sum of simple non-oscillatory exponentially decaying functions in the time domain [3]. For example, the response of a loop of wire (sometimes referred to as a q-coil) may be represented as $r(t) = A \exp(-t/\tau)$ with $\tau = L/R$, where L is the inductance of the loop and R is the resistance. An arbitrarily shaped metallic object will also exhibit an exponentially decaying response with decay rate dependent on the object's shape and constitutive parameters (conductivity and permeability). Therefore, in theory at least, the potential exists to discriminate among metallic objects based on their different decay rates.

Several hardware modifications were made to the AN/PSS-12 to render the device more suitable for discrimination purposes. In particular, we observed that the first stage of amplification following the receiver coil is nonlinear and of limited bandwidth. We therefore bypassed the entire receiver circuitry of the AN/PSS-12 and attached the output of the receiver coil to the input of cascaded AD524 amplifiers—the first (nearest the receive coil) operated at a gain of 10 and the second at a gain of 100. Also a properly adjusted resistance in parallel with the receiver coil yielded a critically damped system response. A National Instruments Scope Card, NI-5102 (15 MHz, 20 MS/s, 8 bits), was used for data acquisition. The data acquisition card resides in the PCMCIA slot of a laptop computer and is controlled using National Instruments LabVIEW software. Our data acquisition system allows the entire exponentially decaying response of a conducting target to be sampled, stored, and then used "off line" to develop discrimination algorithms. An additional increase in system bandwidth was achieved by eliminating the upper half of the original bipolar AN/PSS-12 excitation waveform. The ability to capture an object's entire exponential response provides substantially more information than that available from the original AN/PSS-12 circuitry.

After modifying the AN/PSS-12's transmitter and receiver circuitry as described above, laboratory tests were conducted to ensure that we could capture, with good fidelity, the true exponential response of commonly encountered LM mines. A number of q-coils, all with the same diameter, were constructed using progressively higher wire gauge (thinner wire) so that the exponential decay rate of the loops increased with increasing gauge. The theoretical decay rates of the q-coils were computed and compared with the decay rates extracted from measured q-coil data. Excellent agreement between the two data sets was obtained. We then extracted the decay rates from commonly encountered LM mines and compared their decay rates with those of the q-coils. Because the LM mine decay rates fell within the range of measured loop decay rates, we concluded that our modified AN/PSS-12 data collection system had sufficient bandwidth to accurately measure the decay rate of most LM mines.

Field trials with our modified AN/PSS-12 were conducted at the Fort A. P. Hill, Va., test site. The Joint Unexploded Ordnance Coordination Office in conjunction with the U.S. Army developed this test site, which contains a variety of different mine types and an assortment of metallic clutter commonly encountered in battlefield environments. The test site consists of a large 20 m × 49 m blind test grid and a smaller 5 m × 25 m calibration grid. Every square meter in both grids is set up as a decision opportunity for the detector under test (our modified AN/PSS-12). A landmine or possibly a piece of metallic clutter may be buried at the center of each grid square while some grid squares are intentionally left empty. The calibration grid is used as "ground truth" and data collected there can be used to develop discrimination algorithms. Ground truth associated with the calibration grid is publicly available, whereas only the U.S. government knows ground truth for the blind test grid. Performance is evaluated by an independent government contractor and ultimately presented in terms of a receiver operating characteristic (ROC) curve—created by plotting probability of detection versus probability of false alarm.

Our data collection procedure at Fort A. P. Hill included collecting background data with the search head in contact with the soil. We then subtracted the background data from five measurements made over each target location. The five measurements were made with the target at the center and at the top, bottom, right, and left side of the search coil. Multiple measurements per target location ensure that all independent target modes are measured [6].

A discrimination algorithm based on Bayes' theorem was used to predict, based on the data collected, which squares in the blind grid contained mines. Figure B.2 shows the resulting ROC curve. Note that, at very low false alarm rates, the location of approximately half the mines was correctly identified. The other curves in the figure correspond to algorithms that incorporated only a subset of the five measurements described above. Our research indicates the importance of collecting data that adequately represents all the independent target modes.

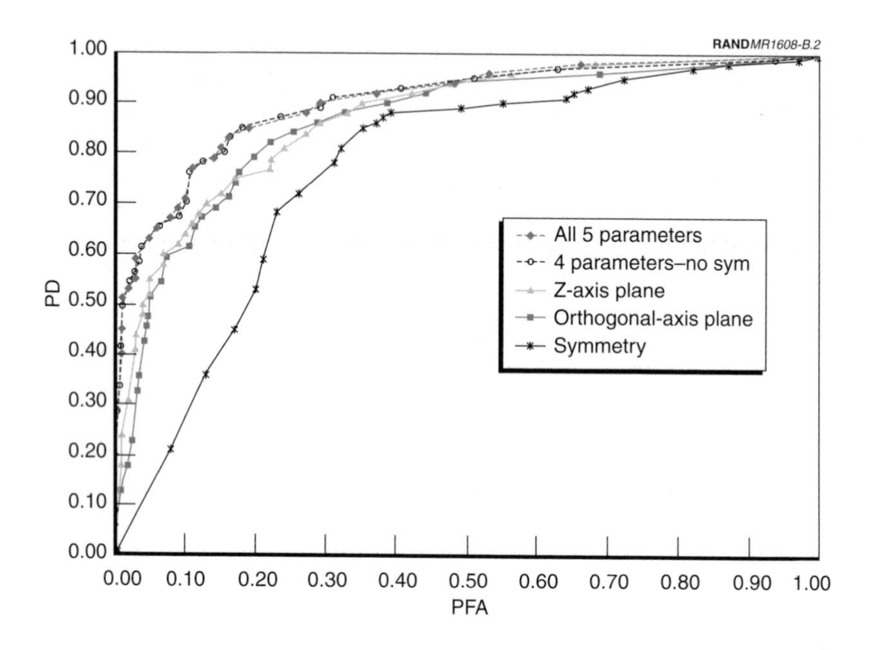


**Figure B.2—ROC Curve Results for Modified AN/PSS-12 Sensor**

## STATE-OF-THE-ART AND FUTURE TRENDS IN EMI DISCRIMINATION RESEARCH

Results in the previous section clearly indicate that the potential exists to discriminate between mines and clutter based on a metallic object's low-frequency quasimagnetostatic response characteristics. It is important to mention that others, in particular researchers at Duke University [7], Johns Hopkins University [8], and AETC Inc. [9], have also conducted successful discrimination experiments. Although discrimination has been successfully demonstrated, the state of the art is still somewhat immature. In particular, "real-time" discrimination capability has yet to be demonstrated and questions remain as to the best way to present the operator with the additional information available from advanced sensors and signal-processing algorithms. A typical envisioned scenario could be that the handheld sensor is operated in two distinct modes. First, the advanced sensor would operate much like a common metal detector, producing an audible alarm whenever a metallic object is nearby. In the second operational mode, perhaps initiated by a simple switch or button, the detector would provide additional information in the form of a heads-up display or electronic voice (or both) indicating which of a set of known threats is most probable. Of course, a host of machine-human interface problems as well as training issues must be addressed before this vision becomes a reality. Improvements in EMI sensor design are also needed. It is desirable to develop an EMI sensor that can pinpoint a target's "center of mass"—and obviously increased sensitivity and bandwidth are also desirable sensor features. Last, environmental issues must also be addressed with particular attention devoted to understanding how highly conducting soils can affect a metallic object's EMI response.

It is the author's opinion that a five-year research and development program funded at approximately $1 million per year could realistically yield a robust real-time handheld discrimination system. The first two years of the envisioned program would focus on sensor improvements, with the next two years devoted to signal-processing development and machine-human interface issues. The final year of the program would be devoted exclusively to field trials leading to final user acceptance.

## REFERENCES

1. L. S. Riggs, "Red Team Final Report: Comparison of the Sensitivity of Five Commonly Used Metal Detectors," Final Report. Submitted to J. Thomas Broach, U.S. Army CECOM RDEC, Night Vision and Electronics Sensors Directorate, ATTN: AMSEL-RD-NV-CD-MD, 10221 Burbeck Road, Ste. 430, Fort Belvoir, VA 22060-5806, March 1999.

2. L. D. Landau and E. M. Lifshitz, *Electrodynamics of Continuous Media,* New York: Pergamon Press, 1960.

3. C. E. Baum, "Low-Frequency Near-Field Magnetic Scattering from Highly, but Not Perfectly, Conducting Bodies," Chapter 6 in *Detection and Identification of Visually Obscured Targets,* C. E. Baum, ed., Philadelphia: Taylor and Francis, 1999.

4. T. Bell, B. Barrow, J. Miller, and D. Keiswetter, "Time and Frequency Domain Electromagnetic Induction Signatures of Unex-

ploded Ordnance," *Subsurface Sensing Technologies and Applications*, Vol. 2, No. 3, 2001, pp. 155–177.

5. L. T. Lowe, L. S. Riggs, S. Cash, and T. Bell, "Improvements to the US Army's AN/PSS-12 Metal Detector to Enhance Landmine Discrimination Capabilities," in *Detection and Remediation Technologies for Mines and Minelike Targets VI*, A. C. Dubey, J. F. Harvey, J. T. Broach, and V. George, eds., Seattle: International Society for Optical Engineering, April 2001, pp. 85–96.

6. C. E. Baum, "The Magnetic Polarizability Dyadic and Point Symmetry," Interaction Note 502, Air Force Weapons Lab, May 1994.

7. L. Collins, P. Gao, and L. Carin, "An Improved Bayesian Decision Theoretic Approach for Land Mine Detection," *IEEE Transactions on Geoscience and Remote Sensing*, Vol. 37, No. 2, March 1998.

8. C. V. Nelson and T. B. Huynh, "Wide Bandwidth, Time Decay Responses from Low-Metal Mines and Ground Voids," in *Detection and Remediation Technologies for Mines and Minelike Targets VI*, A. C. Dubey, J. F. Harvey, J. T. Broach, and R. Dugan, eds., Seattle: International Society for Optical Engineering, April 2001, pp. 55–64.

9. T. H. Bell, B. Barrow, and N. Khadr, "Shape-Based Classification and Discrimination of Subsurface Objects Using Electromagnetic Induction," International Geoscience and Remote Sensing Symposium (IGARSS '98), Seattle, July 1998.

Appendix C

# INFRARED/HYPERSPECTRAL METHODS (PAPER I)

*Brian Baertlein, Ohio State University*

Electro-optical (EO) sensors, which include both infrared bands and hyperspectral sensors, are attractive candidates for some mine detection tasks because they can be used from a considerable standoff distance, they provide information on several mine properties, and they can rapidly survey large areas. Their ability to detect mines has been recognized since the 1950s [1]. These sensors respond to electromagnetic radiation in a sensor-specific wavelength range. The source of the received signal may be either natural (i.e., thermal emission from the target or scattering of sunlight) or artificial (e.g., a laser illuminator), which leads to both active and passive sensor concepts. Antipersonnel mines may be buried or surface-laid. Detection of surface mines, which is a trivial matter for a nearby human observer, is of interest in wide-area search operations using airborne sensors.

In spite of their long history, there is little compelling performance data available for EO detection of antipersonnel mines. Two likely reasons for this are as follows: First, EO mine signatures tend to be highly dependent on environmental conditions, which complicates data collection and performance assessment. Indeed, a number of multisensor mine detection systems have included an EO sensor (typically, a thermal imager) but have not used it because of unreliable data quality. As a result, much of the work to date comprises concept demonstrations and studies of specific phenomena. Second, the only EO performance tests involving statistically meaningful sample sizes have been conducted on U.S. Army test sites containing primarily antitank mines. Detection of antipersonnel mines is significantly more challenging than detection of antitank mines, and performance results for the two cases are not easily related. Furthermore, much of the recent work addresses detection of surface-laid antitank minefields. Minefield detection is typically much easier than detection of single mines because of the large number of mines involved and the additional spatial information (e.g., the presence of a spatial pattern). Some data have been acquired at mixed antipersonnel/antitank test sites in the United States, but scores for the two mine classes are typically not reported separately. Significant European research in antipersonnel mine detection is under way, but test sites commonly used in those efforts [2,3] are smaller.

Both thermal emissions and surface-scattering phenomena contribute to EO mine signatures, and the relative importance of these phenomena depends on whether the mine is deployed on the surface or is buried. The discussion that follows is organized first by mine deployment, and then by detection approach. References to relevant performance data are provided when available, as are specific limitations. Recommendations for research and development (R&D) programs that span several sensor concepts are presented at the end of this appendix.

## BURIED MINES

Detection of buried antipersonnel mines comprises a very challenging problem. A number of concepts have been examined for this problem.

### Passive Thermal Detection

**Physical Basis.** A large part of the solar energy incident on soil is absorbed, leading to heating. As a result of this heating, the soil emits thermal radiation detectable by a thermal infrared (IR) sensor. Natural solar heating and cooling over a diurnal cycle tend to affect a buried object and the surrounding soil differently, which leads to a detectable temperature difference. For a buried mine this difference arises because the mine is a better thermal insulator than the soil. During the day, the thin layer of soil over the mine tends to accumulate thermal energy because the mine impedes the transport of that heat deeper into the ground. As a result, soil over a mine will tend to be warmer than the surrounding soil. Conversely, in the evening hours, the soil layer over the mine gives up its thermal energy more rapidly than the surrounding soil and it appears cooler. Twice daily the soil over the mine and the background soil will assume the same temperature, making thermal detection impossible. The temperature difference and its temporal behavior depend strongly on a variety of variable natural phenomena, including the time of day, prior solar illumination, wind speed, ground cover, and soil composition (e.g., moisture content).

Most thermal detection concepts involve single looks ("snapshots") of the region of interest. The soil over a mine has different thermal dynamics than homogeneous soil and, as a result, a time sequence of images can often produce better detection than a single image. Hence, staring sensors, which are impractical for many military scenarios, may be attractive for humanitarian demining.

**Development Status.** Broadband passive sensors at IR wavelengths are mature and available commercially from several vendors. Algorithms for mine detection, another critical part of any detection system, are somewhat less mature, although a number of groups have reported progress in his area.

**Current Performance.** As noted, compelling performance data for EO detection of antipersonnel mines are limited. Receiver operating characteristic (ROC) curves for thermal IR detection of a mixture of antipersonnel and antitank mines were reported by Baertlein and Gunatilaka [4], but those results comprised only 27 mines (examined under poor environmental conditions), of which roughly half were antipersonnel. Data from the TNO (Netherlands Organisation for Applied Scientific Research) mine lanes in the Netherlands were examined by Milisavljevic et al. [5], and performance results are reported therein for 15 antipersonnel mines. Additional results involving 18 antipersonnel mines were reported by Chen et al. [6]. In that work the data collection extended over several hours, and temporal processing was used to improve performance. Limited results on thermal IR detection of buried antitank minefields from airborne sensors are summarized in Miles, Cespedes, and Goodson [17].

**Limitations.** Signature variations with time and environmental conditions are a persistent problem for thermal IR mine detection. The optimum time for detection and the expected contrast depend on factors noted above that are often unknown to a remote observer. Surface clutter from reflected light and inhomogeneous soil properties are also problematic. In many cases the size of these clutter artifacts is comparable to that of antipersonnel mines, which leads to false alarms. Thermal emission from foliage (at the temperature of living, respiring vegetation) tends to mask the temperature of the underlying soil (and the thermal mine signature).

**Potential for Improvement.** The processes that produce thermal IR target signatures and clutter are poorly understood. In some cases, good detection performance has been demonstrated, but when such systems fail, the reasons for failure are often not evident. A better understanding of target and clutter signatures could substantially improve their effectiveness by allowing them to be deployed appropriately. Staring sensors should also be considered, which take data over an extended period in time, waiting for favorable conditions to arise. Time-history information will also help to compensate for the variability of thermal signatures with time and environmental conditions.

## Active Thermal Detection

**Physical Basis.** Passive thermal detection is based on solar heating of the soil, and it is prone to fail when environmental conditions are not conducive. Active analogs of the process have been investigated, in which intense optical [7,8] or high-power microwave (HPM) [9–11] sources are used. Soil has a low optical albedo and a moderately high radio frequency conductivity, both of which lead to effective heating by external sources. In contrast, the mine is typically either plastic (a good electrical and thermal insulator) or metallic (a good electrical and thermal conductor). The HPM approach to heating is particularly attractive because an HPM antenna can be shared by a ground-penetrating radar (GPR). In addition to the thermal effects of HPM, the presence of the mine manifests itself in another way: The mine's dielectric discontinuity produces reflections of the illuminating microwave field (a "standing wave pattern"), which affects HPM absorption and heating.

**Development Status.** The investigation of this approach has not progressed beyond small-scale experiments. Progress is not significantly hampered by the instrumentation. Suitable optical and HPM sources are available, and the thermal IR sensor required here is identical to that described in the previous paragraph.

**Current Performance.** Controlled demonstrations using heat lamps [7], lasers [8], microwaves [9,10], and two-frequency microwaves [11] have been presented, but only a few mines were imaged. No performance data are available.

**Limitations.** Relatively long exposures (up to 12 minutes in Storm and Haugsted [7]) are required to heat the soil, and the peak contrast may not be observed for several minutes after heating, but suitable operational concepts could be defined to make this detection paradigm practical. The HPM approach is somewhat better developed than the optical approaches. The key issues are (1) producing uniform illumination on a rough ground, (2) producing sufficient power to heat the ground at a distance, and (3) avoiding the human health hazards associated with HPM.

**Potential for Improvement.** The ability to generate a new sensor paradigm by simply adding an HPM source to an existing GPR system is attractive. Studies of this detection concept are incomplete, but the dynamics of the thermal processes suggest that it is impractical for rapid area scans. Additional research will be required to determine the limits of the method.

## Passive Detection of Nonthermal Surface Phenomena

**Physical Basis.** Buried mines are also detectable via soil disturbances and vegetation stress. Soil comprises a mixture of materials having a range of particle sizes. Natural processes tend to move the smaller particles deeper into the soil. Excavation for mine burial brings the smaller particles to the surface, where they affect surface scattering. The most effective sensors of this behavior are hyperspectral, although polarimetric effects have also been alleged. Because it obscures the surface, vegetation presents additional challenges to buried mine detection, but another phenomenon can be exploited in this case. The mine presents a moisture barrier to the upward and downward flow of soil water. This leads to a (temporary) pooling of water over the top of the mine after a period of rain and drier soil over the mine in the absence of rain. The latter condition tends to produce drought stress on the vegetation, which can be detected with a hyperspectral sensor.

**Development Status.** A large-scale experimental collection of the underlying hyperspectral signatures (0.35–14 μm) for soil and buried mines was performed by Veridian-ERIM [12], with a subsequent analysis by Kenton et al. [13]. An outdoor surface mine collection using a nonimaging spectrometer was reported by Haskett et al. [14]. A hyperspectral imaging visible/near infrared (VNIR) sensor has been flown on an airborne platform by the Canadian Defence Research Establishment–Suffield (DRES) [15]. The U.S. Marine Corps developed the Coastal Battlefield Reconnaissance and Analysis (COBRA) system [16], which employs a multispectral camera (VNIR) on an airborne platform.

**Current Performance.** Performance data for 18 mines buried under three soil types are reported by Haskett et al. [14]. Limited performance data are reported in McFee and Ripley [15] for buried antitank and antipersonnel mines under a variety of ground covers. Those data were acquired after the mines had been in place for some time. The antipersonnel mine detection performance was encouraging, but the sensor was hampered by insufficient resolution for these smaller mines. Detection of minefields, portions of which were underwater, is discussed in Stetson et al. [16] for the COBRA system.

**Limitations.** A number of challenges are encountered in nonthermal detection of buried mines. The effect of the soil disturbances described above is transient and is greatly reduced by rainfall. Vegetation stress also depends on recent rainfall, but it is also a longer-term effect, which may not be evident for recently buried mines. Both phenomena are unreliable in areas with broken grass or low shrubs, which present false alarms.

**Potential for Improvement.** The findings of McFee and Ripley [15] bear further study. A hyperspectral sensor with high spatial resolution should be developed for antipersonnel mine detection.

## Active Sensing of Nonthermal Surface Phenomena

**Physical Basis.** Active hyperspectral or polarimetric sensing of the phenomena described in the previous section are also feasible for buried mines. This approach circumvents the problem of uncontrolled, variable solar illumination by using a scanned laser illuminator. An accompanying narrowband receiver can be used to reject ambient light, thereby improving image contrast.

**Development Status.** From 1987 to 1992, the U.S. Army developed the Remote Minefield Detection System (REMIDS) [17] as part of the Standoff Minefield Detection System (STAMIDS). REMIDS comprises an airborne sensor package using both active polarimetric and passive thermal sensors. The system was used in field tests at Fort Hunter-Liggett, Calif., and Fort Drum, N.Y., in 1990–1991. This technology formed the basis for the Airborne STAMIDS (ASTAMIDS) program [18], which was field-tested during 1996.

**Current Performance.** The REMIDS sensor has been tested against both surface-laid and buried mines. A summary of the performance against antitank minefields is presented in Miles, Cespedes, and Goodson [17]. Limited results from the ASTAMIDS sensor are summarized in [19] for buried and surface-laid antitank minefields. Buried mine detection performance was limited.

**Limitations.** Some relevant issues are noted in the previous Limitations subsection. The detection range of active sensors is also necessarily limited by the transmitter power.

**Potential for Improvement.** Prior work with active polarimetric sensors (REMIDS) for buried antitank mines has been disappointing. To date, there does not appear to have been a study of hyperspectral sensors for this application. The modest successes described above in Passive Detection (under Current Performance) should be explored for active detection of antipersonnel mines using a suitable spatial resolution.

## SURFACE MINES

As noted, surface mine detection is primarily of interest for airborne or other platforms with an appreciable standoff distance. At large distances, the number of pixels on the mine decreases, but some compensating factors exist. First, the surface scattering properties of the mine are detectable in addition to thermal phenomena. Second, whereas the thermal signatures of buried mines often have indistinct shapes, the shapes of surface mine signatures contain significant information and may be useful discriminators. Finally, when the illumination arrives at low elevation angles, shadows may also be exploited in detection.

## Passive Thermal Detection

**Physical Basis.** Because a mine's thermal properties are considerably different than those of vegetation, a solar-heated mine viewed with a thermal IR sensor typically has a high contrast. This contrast often exists even when the mine is painted to camouflage its presence. Differences in paints or coloration on different parts of the mine (e.g., the central trigger assembly vis-à-vis the main body) may lead to complex, distinctive thermal signatures because of different solar absorptions.

**Development Status.** The U.S. Department of Defense has developed several IR minefield detection systems for airborne and vehicle platforms, including the REMIDS and ASTAMIDS systems. As noted previously, those systems have been principally used against antitank mines. Another more recent development is the U.S. Army Lightweight Airborne Minefield Detection (LAMD) system, in which both passive thermal IR and active polarimetric near infrared (NIR) sensors will be used against surface antitank mines. The development of an interim system is reported in Trang [20].

**Current Performance.** Information on thermal IR detection performance for surface-laid antipersonnel mines is not available in the literature, but there are significant data on antitank mine detection. The performances of some airborne thermal sensors of surface minefields are reported in Stetson et al. [16]. ROC curves for detection of surface antitank mines using a thermal IR sensor were reported in selected reports [20–22].

**Limitations.** With due attention to the increased role of surface scattering and the more rapid time-dependence of surface mine heating, the limitations noted above for buried mine detection are also relevant here.

**Potential for Improvement.** As noted for buried mines, there is a need to better understand the signatures and their relation to environmental conditions. Knowledge of the sun angle could be used to predict heating patterns, which may improve detection and false alarm rejection.

## Passive Nonthermal Detection

**Physical Basis.** The surface scattering properties of mines are distinct from those of soil and vegetation, particularly when measured in the spectral domain, and that spectral dependence can be a powerful discriminator. In addition to their spectral properties, many mines are relatively flat and covered with the same material over much of their top surface. This leads to the appearance of a uniform region in the imagery, which tends to be useful in image processing. The polarization properties of surface-laid mines can also be exploited by a passive sensor. The polarimetric signature of unstructured random surfaces such as grass tends to be random itself, which leads to an unpolarized return. In contrast, the smooth surfaces of man-made materials tend to produce a polarized signature when viewed at low elevation angles. Polarimetric signatures can exist even where there is no detectable thermal signature.

**Development Status.** Detectors of passive broadband polarimetric emissions have been described by the Defence Evaluation and Research Agency (DERA) in the United Kingdom [23], by Larive et al. in France [24], and by Cremer et al. in the Netherlands [25], all of which are based on commercial mid-wave infrared (MWIR) cameras. The DERA system, developed by Nichols Research in the United States, uses a micropolarizer array bonded to the focal plane array. The French and TNO systems use uncooled rotating wire-grid polarizers in front of the sensor. The DERA system has been used in vehicle-mounted data collections, and work on the data processing algorithms has been described. The TNO system has been used for tests over a diurnal cycle while staring at emplaced mine surrogates. Significant progress was reported in modeling the signatures. A passive imaging, hyperspectral, polarimetric sensor has been demonstrated by Iannarilli et al. [26]. The sensor was used to image mine surrogates, and techniques for data analysis were presented. Airborne detection of surface minefields has been demonstrated by the DRES hyperspectral system [27]. That test involved approximately half antitank and half antipersonnel mines.

**Current Performance.** Few detection performance estimates are available for passive nonthermal detection of antipersonnel mines. DRES [27] reported the performance of ground-based and airborne hyperspectral sensors using several detection algorithms. Limited antitank mine detection performance estimates for the multispectral COBRA sensor are available in Stetson et al. [16].

**Limitations.** The sensor concepts described here, like the other passive sensors described above, are limited by uncontrollable variation in solar illumination. Surface mine signatures are also strongly sensitive to the sun angle, which causes shadows and heating on specific parts of the mine. Additional sensor-specific issues arise. Passive polarimetric signatures tend to be weak for mines with rough surfaces. For mines with smooth surfaces, the polarimetric signature is strongest when viewed at low elevation angles. Unfortunately, at those angles, the mines can also be obscured by vegetation. Hyperspectral sensors also have limits. In principle, the spectral signature of mines can be measured and subsequently used to improve detection algorithms, but the unpredictable effects of rust, dirt, and material aging make it difficult to do so. Fielded hyperspectral detection algorithms often comprise simple anomaly detectors.

**Potential for Improvement.** Greater use of hyperspectral and polarimetric methods will permit more information per pixel, which aids detection. More extensive tests should be conducted on antipersonnel mines to determine the true performance of these sensors. In such work, improved spatial resolution will be required. Finally improvements in image processing techniques are likely to offer significant gains.

## Active Sensing for Surface Mines

**Physical Basis.** Only active sensors of nonthermal phenomena are of interest because actively provoking a thermal signature from a significant distance requires impractical amounts of power. The relevant physical phenomena for this sensor concept have been described above. The use of active sensors is appealing for polarimetric sensors in which a fixed polarization can be transmitted. The basis for an active polarimetric sensor is quite different from the passive polarimetric sensors noted above, in which low elevation angles are preferred to detect the polarized signature of the mine. Active sensors tend to operate at near-nadir viewing angles, where a smooth mine surface will not depolarize the (polarized) illumination, while scattering from randomly oriented foliage will be depolarized.

**Development Status.** Active sensors have been extensively investigated by a number of U.S. programs, including the REMIDS and ASTAMIDS studies. The Army is currently developing the LAMD-Laser system, in which a polarized laser illuminator will be used to detect co- and cross-polarized returns from surface antitank mines. Current plans call for fusion of the active sensor and a passive MWIR imager. Other sensors have been investigated. Three active sensors operating in the NIR and short-wave infrared (SWIR) bands have been demonstrated by de Jong et al. [28] in the Netherlands. Those experimental sensors were used in controlled, small-scale outdoor tests on a smooth sand background. An active hyperspectral imaging system was demonstrated by Johnson et al. [29], who imaged a number of artificial materials in vegetation. Tripwire detection is a closely related field of interest. Relevant work is described by Allik et al. [30] for SWIR sensors and by Babey et al. [31] for ultraviolet, VNIR, and SWIR.

**Current Performance.** Results were presented in de Jong, Winkel, and Roos [28] for a test site having 25 surface-laid mines and clutter objects on a smooth sand background. Although most of the objects show high contrast, no performance data are provided. Summary minefield performance data for the REMIDS sensor are given in Stetson et al. [16]. No performance data are yet available for the LAMD-Laser system, although at the time of this writing preliminary data had been acquired from several tests sites. Sensor enhancements and algorithm development are under way.

**Limitations.** A number of relevant issues were noted in the previous Limitations subsection. The need to operate at near-nadir viewing angles for active polarimetric sensors was described above.

**Potential for Improvement.** As noted, there is little performance data available on antipersonnel mines for active sensors, but exploratory tests have been encouraging. Future work should focus on a quantitative assessment of this sensor concept.

## R&D PROGRAM RECOMMENDATIONS

EO sensors are among the most attractive technologies for achieving wide-area antipersonnel mine detection. Although they have been investigated at some length for antitank mines (and minefields), they have not been explored significantly for antipersonnel mines. To address this deficiency, any R&D program should include the following activities:

- creation of an antipersonnel mine test range with a statistically significant number of mines
- collection of baseline data sets on that test range
- development of suitable processing algorithms leading to performance statistics
- closer collaboration with European colleagues working in this area.

Some specific research goals are as follows:

### Passive Thermal Detection of Buried and Surface-Laid Mines

A better understanding of the processes that produce thermal mine signatures should be initiated for both buried and surface mines. A combination of theoretical and experimental work is required. The environmental parameters required to predict thermal IR performance should be determined. The goal of the work should be a model using environmental parameters as inputs and capable of predicting the best times to deploy thermal IR sensors.

### Active Thermal Detection of Buried Mines

Further investigation of the HPM approach is warranted, beginning with an investigation of the nonuniform illumination problem and definition of an operational sensor concept.

## Active and Passive Hyperspectral Detection of Mines

Buried mine detection should be investigated first. A better understanding of the phenomena that produce hyperspectral signatures for disturbed soil and vegetation stress should be undertaken. Signatures should be acquired both for recently buried mines and for mines that have been allowed to "weather in." Both bare soil and vegetated surfaces should be examined. A hyperspectral sensor with high spatial and spectral resolution should be developed and used in the testing. The sensor should also be used to image both buried and surface mines, and a comparison of hyperspectral and polarimetric detection (see next paragraph) should be undertaken.

## Active and Passive Polarimetric Detection of Surface Mines

Past experience with active polarimetric sensors (e.g., REMIDS) for surface antitank minefields has demonstrated encouraging performance, and further development of this concept (e.g., LAMD-Laser) is under way. Tests of new sensors should also be performed for antipersonnel mines using commensurately smaller spatial resolution. A passive, imaging polarimetric sensor should be fielded and tested. Data on a large set of antipersonnel mines should be acquired. The data acquisition effort should be supported by a parallel effort in detection algorithm development.

## REFERENCES

1. C. Stewart, *Summary of Mine Detection Research*, Vol. I., Technical Report 1636-TR, Fort Belvoir, Va.: U.S. Army Engineer Research and Development Laboratories (now NVESD), Corps of Engineers, 1960.

2. W. de Jong, H. A. Lensen, and Y. H. L. Janssen, "Sophisticated Test Facility to Detect Landmines," in *Detection and Remediation Technologies for Mines and Minelike Targets IV*, A. C. Dubey, J. F. Harvey, J. Broach, and R. E. Dugan, eds., Seattle: International Society for Optical Engineering, 1999, pp. 1409–1418.

3. P. Verlinde, M. Acheroy, G. Nesti, and A. Sieber, "First Results of the Joint Multi-sensor Mine-signatures Measurement Campaign

(MsMs Project)," in *Detection and Remediation Technologies for Mines and Minelike Targets VI*, A. C. Dubey, J. F. Harvey, J. T. Broach, and V. George, eds., Seattle: International Society for Optical Engineering, 2001, pp. 1023–1034.

4. B. A. Baertlein and A. H. Gunatilaka, "Optimizing Fusion Architecture for Limited Training Data," in *Detection and Remediation Technologies for Mines and Minelike Targets V*, A. C. Dubey, J. F. Harvey, J. Broach, and R. E. Dugan, eds., Seattle: International Society for Optical Engineering, 2000, pp. 804–815.

5. N. Milisavljevic, S. P. van den Broek, I. Bloch, P. B. W. Schwering, H. A. Lensen, and M. Acheroy, "Comparison of Belief Functions and Voting Methods for Fusion of Mine Detection Sensors," in *Detection and Remediation Technologies for Mines and Minelike Targets VI*, A. C. Dubey, J. F. Harvey, J. T. Broach, and V. George, eds., Seattle: International Society for Optical Engineering, 2001, pp. 1011–1022.

6. D. H. Chen, I. K. Sendur, W. J. Liao, and B. A. Baertlein, "Using Physical Models to Improve Thermal IR Detection of Buried Mines," in *Detection and Remediation Technologies for Mines and Minelike Targets VI*, A. C. Dubey, J. F. Harvey, J. T. Broach, and V. George, eds., Seattle: International Society for Optical Engineering, 2001, pp. 207–218.

7. J. Storm and B. Haugsted, "Detection of Buried Land Mines Facilitated by Actively Provoked IR Signatures," in *Detection and Remediation Technologies for Mines and Minelike Targets IV*, A. C. Dubey, J. F. Harvey, J. T. Broach, and R. Dugan, eds., Seattle: International Society for Optical Engineering, 1999, pp. 167–172.

8. D. E. Poulain, S. A. Schuab, D. R. Alexander, and J. K. Krause, "Detection and Location of Buried Objects Using Active Thermal Sensing," in *Detection and Remediation Technologies for Mines and Minelike Targets III*, A. C. Dubey, J. F. Harvey, and J. Broach, eds., Seattle: International Society for Optical Engineering, 1998, pp. 861–866.

9. C. A. DiMarzio, C. M. Rappaport, and L. Wen, "Microwave-Enhanced Infrared Thermography," in *Detection and Remediation Technologies for Mines and Minelike Targets III*, A. C. Dubey,

J. F. Harvey, and J. Broach, eds., Seattle: International Society for Optical Engineering, 1998, pp. 1103–1110.

10. S. M. Khanna, F. Paquet, R. Apps, and J. S. Sereglyi, "New Hybrid Remote Sensing Method Using HPM Illumination/IR Detection for Mine Detection," in *Detection and Remediation Technologies for Mines and Minelike Targets III*, A. C. Dubey, J. F. Harvey, and J. Broach, eds., Seattle: International Society for Optical Engineering, 1998, pp. 1111–1121.

11. T. Shi, G. O. Sauermann, C. M. Rappaport, and C. A. DiMarzio, "Dual-Frequency Microwave-Enhanced Infrared Thermography," in *Detection and Remediation Technologies for Mines and Minelike Targets VI*, A. C. Dubey, J. F. Harvey, J. T. Broach, V. George, eds., Seattle: International Society for Optical Engineering, 2001, pp. 379–397.

12. A. M. Smith, R. Horvath, L. S. Nooden, J. A. Wright, J. L. Michael, and A. C. Kenton, *Mine Spectral Signature Collections and Data Archive*, Technical Report #10012200-15-T under project DAAB07-98-C-6009, CDRL A004, ERIM International Inc., Ann Arbor, Mich., March 1999.

13. A. C. Kenton, W. A. Malila, L. S. Nooden, V. E. Diehl, and K. Montavon, "Joint Spectral Region Buried Land Mine Discrimination Performance," in *Detection and Remediation Technologies for Mines and Minelike Targets V*, A. C. Dubey, J. F. Harvey, J. T. Broach, and R. Dugan, eds., Seattle: International Society for Optical Engineering, 2000, pp. 210–219.

14. H. T. Haskett, D. A. Reago, and R. R. Rupp, "Detectability of Buried Mines in 3–4 μm and 8–12 μm Regions for Various Soils Using Hyperspectral Signatures," in *Detection and Remediation Technologies for Mines and Minelike Targets VI*, A. C. Dubey, J. F. Harvey, J. T. Broach, and V. George, eds., Seattle: International Society for Optical Engineering, 2001, pp. 296–309.

15. J. E. McFee and H. T. Ripley, "Detection of Buried Landmines Using a Case Hyperspectral Imager," in *Detection and Remediation Technologies for Mines and Minelike Targets II*, A. C. Dubey and R. L. Barnard, eds., Seattle: International Society for Optical Engineering, 1997, pp. 738–749.

16. S. P. Stetson, N. H. Witherspoon, J. H. Holloway, Jr., and H. R. Suiter, "COBRA ATD Minefield Detection Results for the Joint Countermine ACTD Demonstrations," in *Detection and Remediation Technologies for Mines and Minelike Targets V,* A. C. Dubey, J. F. Harvey, J. Broach, and R. E. Dugan, eds., Seattle: International Society for Optical Engineering, 2000, pp. 1268–1279.

17. B. H. Miles, E. R. Cespedes, and R. A. Goodson, "Polarization-Based Active/Passive Scanning System for Minefield Detection," in *Polarization and Remote Sensing,* Seattle: International Society for Optical Engineering, 1992, pp. 239–252.

18. G. Maksymonko and K. Brieter, "ASTAMIDS Minefield Detection Performance at Aberdeen Proving Ground Test Site," in *Detection and Remediation Technologies for Mines and Minelike Targets II,* A. C. Dubey and R. L. Barnard, eds., Seattle: International Society for Optical Engineering, 1997, pp. 726–737.

19. J. G. Ackenhusen, Q. A. Holmes, C. King, and J. A. Wright, *State-of-the-Art Report on Detection of Mines and Minefields,* Technical Report (Draft), Fort Belvoir, Va.: Infrared Information Analysis Center, U.S. Army Night Vision and Electronic Sensors Directorate, May 31, 2001.

20. A. Trang, *Lightweight Airborne Multispectral Minefield Detection—Interim (LAMD-I): Final Report,* Fort Belvoir, Va.: U.S. Army Night Vision Electronic Systems Directorate, 2002.

21. B. A. Baertlein, *Modeling and Signal Processing Support to the Lightweight Airborne Multispectral Minefield Detection (LAMD-I) Program: Final Report,* Columbus, Ohio: Ohio State University ElectroScience Laboratory, Technical Report, 739345-1, 2002.

22. S. Agarwal, P. Sriram, P. P. Palit, and O. R. Mitchell, "Algorithms for IR Imagery Based Airborne Landmine and Minefield Detection," in *Detection and Remediation Technologies for Mines and Minelike Targets VI,* A. C. Dubey, J. F. Harvey, J. T. Broach, and V. George, eds., Seattle: International Society for Optical Engineering, 2001, pp. 284–295.

23. J. W. Williams, H. S. Tee, and M. A. Poulter, "Infrared Processing and Classification for the UK Remote Minefield Detection System Infrared Polarimetric Camera," in *Detection and Remedia-*

*tion Technologies for Mines and Minelike Targets VI*, A. C. Dubey, J. F. Harvey, J. T. Broach, and V. George, eds., Seattle: International Society for Optical Engineering, 2001, pp. 139–152.

24. M. Larive, L. Collot, S. Breugnot, H. Botma, and P. Roos, "Laid and Flush Buried Mine Detection Using 8–12 μm Polarimetric Imager," in *Detection and Remediation Technologies for Mines and Minelike Targets III*, A. C. Dubey, J. F. Harvey, and J. Broach, eds., Seattle: International Society for Optical Engineering, 1998, pp. 115–120.

25. F. Cremer, W. de Jong, K. Schutte, J. T. Johnson, and B. A. Baertlein, "Surface Mine Signature Modeling for Passive Polarimetric IR," in *Detection and Remediation Technologies for Mines and Minelike Targets VII*, J. T. Broach, R. S. Harmon, and G. J. Dobek, eds., Seattle: International Society for Optical Engineering, 2002, pp. 51–62.

26. F. J. Iannarilli, Jr., H. E. Scott, and S. H. Jones, "Passive IR Polarimetric Hyperspectral Imaging Contributions to Multi-Sensor Humanitarian Demining," in *Detection and Remediation Technologies for Mines and Minelike Targets VI*, A. C. Dubey, J. F. Harvey, J. T. Broach, and V. George, eds., Seattle: International Society for Optical Engineering, 2001, pp. 348–352.

27. S. B. Achal, C. D. Anger, J. E. McFee, and R. W. Herring, "Detection of Surface-Laid Mine Fields in VNIR Hyperspectral High Spatial Resolution Data," in *Detection and Remediation Technologies for Mines and Minelike Targets IV*, A. C. Dubey, J. F. Harvey, J. Broach, and R. E. Dugan, eds., Seattle: International Society for Optical Engineering, 1999, pp. 808–818.

28. A. N. de Jong, H. Winkel, and M. J. J. Roos, "Active Multispectral Near-IR Detection of Small Surface Targets," in *Detection and Remediation Technologies for Mines and Minelike Targets VI*, A. C. Dubey, J. F. Harvey, J. T. Broach, and V. George, eds., Seattle: International Society for Optical Engineering, 2001, pp. 353–364.

29. B. Johnson, R. Joseph, M. Nischan, A. Newbury, J. Kerekes, H. Barclay, B. Willard, and J. Zayhowski, "A Compact, Active Hyperspectral Imaging System for the Detection of Concealed Targets," in *Detection and Remediation Technologies for Mines and Mine-*

*like Targets IV*, A. C. Dubey, J. F. Harvey, J. Broach, and R. E. Dugan, eds., Seattle: International Society for Optical Engineering, 1999, pp. 144–153.

30. T. Allik, D. Dawkins, J. Habersat, J. Nettleton, T. Writer, C. Buckingham, H. Schoeberlein, C. V. Nelson, and K. Sherbondy, "Eyesafe Laser Illuminated Tripwire (ELIT) Detector," in *Detection and Remediation Technologies for Mines and Minelike Targets VI*, A. C. Dubey, J. F. Harvey, J. T. Broach, and V. George, eds., Seattle: International Society for Optical Engineering, 2001, pp. 176–186.

31. S. K. Babey, J. E. McFee, C. D. Anger, A. Moise, and S. B. Achal, "Feasibility of Optical Detection of Landmine Tripwires," in *Detection and Remediation Technologies for Mines and Minelike Targets V*, A. C. Dubey, J. F. Harvey, J. Broach, and R. E. Dugan, eds., Seattle: International Society for Optical Engineering, 2000, pp. 220–231.

Appendix D

# INFRARED/HYPERSPECTRAL METHODS (PAPER II)

*John G. Ackenhusen, Veridian International*[1]

## BASIC PHYSICAL PRINCIPLES

This appendix addresses landmine detection performance for infrared (IR) and hyperspectral (HS) sensors, which include broadband, multispectral (2–20 bands), and hyperspectral (more than 20 bands). Bands include the VNIR, SWIR, MWIR (both reflective and thermal), and LWIR.[2] It also considers the use of polarization information within these bands.

Figure D.1 displays the myriad of conditions under which mine detection must be accomplished. Indeed, one of the factors determining the success of a sensor is its ability to cover this variety of conditions within the environment, ground cover, soil, and mine.

Broadband IR mine detection operates in the thermal domain, measuring the apparent temperature difference between the target and background. Here, the target can be either a mine on the surface of the ground or the area over a buried mine. The presence of a mine leads to a difference in the rate of heating or cooling of the area over the mine, producing a diurnal cycle to the signature and a contrast reversal of the mine area versus the background that occurs over the daily cycle.

---

[1]The author acknowledges the assistance of Jack Cederquist, Robert Horvath, and Craig Kenton in preparing this work. Originally published by Veridian International, 2001. Reprinted with permission.

[2]VNIR, SWIR, MWIR, and LWIR are visible/near, short-wave, mid-wave, and long-wave infrared, respectively.

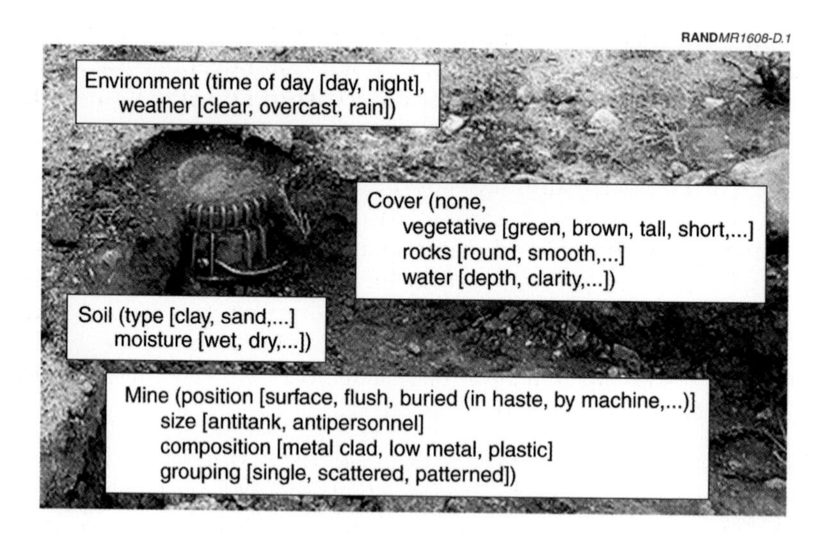


**Figure D.1—Successful IR/HS Mine Detection Must Accommodate a Wide Variety of Conditions**

Spectral mine detection either examines the apparent temperature difference in more detail (because multiple bands are used instead of one) or detects reflective color difference of mines or their covering material with respect to their backgrounds. Spectral detection depends on detecting the effect on soil or vegetation as a result of burying the mine, not the buried mine itself. Upon burying the mine in bare soil, the placement or presence of the mine will change the observables of a small region around it. Immediately upon placement, particle size, texture, or moisture differences can be detected by broadband IR or spectral methods. However, most prominent effects weather away with time.

The largest bare-soil effect is the result of a change in the distribution of particle sizes of the soil upon disturbance for mine burial. Soils consist of a range of particle sizes, and small particles (e.g., 2 μm) are much more mobile than larger particles. Spectral behavior depends on the soil particle size relative to the wavelength of light. Freshly disturbed soil has more fine particles that cover larger particles. This covering effect suppresses or smooths out the spectral features of the large particles in recently disturbed areas. Undisturbed areas tend to display spectral characteristics of the bulk (larger particles) of the soil. Examples include the quartz doublet feature centered at 8.5 μm and 8.9 μm—fine particles of recently disturbed soil suppress this feature, but undisturbed soils display it more strongly.

Other spectral measurements detecting disturbance are based on color changes in vegetation—VNIR and SWIR spectra can show change in chlorophyll and water spectral features. Disturbances of soils high in organic materials may be detected in SWIR by observing features from lignin and cellulose. Mine burial may permanently alter the vigor of overlying vegetation if the root zone is greatly disturbed. Excavating other minerals during burial may also provide spectral feature evidence of disturbance.

Polarimetric effects are useful for detecting surface mines. Passive polarimetry either uses the sun or sky for illumination, or uses thermal emissions. Active polarization uses a probe light, generally from a laser, and looks at polarization of the returned light. Natural materials tend to depolarize the radiation that is returned, while man-made materials, which are smoother, tend to preserve the polarization of incident radiation upon reflection, attributed to specular surfaces, or emit polarized light in accordance with the Fresnel equations. By supplying the illumination, active polarization establishes a known geometry (position and polarization of the light source) and is more reliable than passive polarization (for which the light source, such as the sun, varies in location).

## STATE OF DEVELOPMENT

The maturity of IR/HS mine detection techniques is determined more by algorithm capability (i.e., the ability to transform the observed sensor outputs into decisions on the presence or absence of mines or minefields) rather than sensor maturity. All three techniques (broadband, spectral, and polarimetric) have been the subject of recent field tests. These are considered in order of decreasing maturity.

Work in broadband IR detection of mines has been conducted by Northrop Grumman, TRW, BAE Systems, and the government organizations of the U.S. Army Night Vision and Electronic Sensors Directorate (NVESD) and the UK Defence Evaluation and Research

Agency (DERA). Broadband IR sensors are commercially available, and such sensors have been used in handheld, vehicle-based, and airborne field tests. However, the processing algorithms have not been able to detect mines with sufficient accuracy, or to discriminate mines from other mine-like objects, to meet requirements.

Spectral mine detection has been carried out by companies that include Aerospace Corp., BAE Systems, Raytheon, Space Computer Corporation, SAIC, the University of Hawaii, and Veridian. Active government organizations include the Defense Advanced Research Projects Agency (DARPA), the Navy's Coastal Systems Station (CSS), and NVESD.

The market for spectral sensors is extremely limited, becoming smaller as both the number of bands and the wavelength increase. Commercial off-the-shelf sensors are restricted to multispectral sensors in the VNIR range, such as those sold by Xybion. As the wavelength increases to LWIR, the technology becomes more complex, requiring more expensive focal plane detector arrays and cooling. However, successful one-of-a-kind hyperspectral LWIR sensors have been built, and some current U.S. government programs are directed to building LWIR hyperspectral sensors with a form factor compatible with unmanned aerial vehicles (e.g., BAE Systems, Long Island, N.Y.; Raytheon, Plano, Tex.). VNIR multispectral minefield detection has been proven to work in the littoral zone under the Coastal Battlefield Reconnaissance and Analysis (COBRA) program of CSS and Veridian, meeting the operational minefield detection requirements set by this advanced technology demonstration. The basic phenomenology of hyperspectral minefield detection has been studied to identify those observables that may be sensed to detect mines (NVESD, Veridian, MTL, and SAIC). STI is now flying a prototype visible HS sensor ("LASH-MCM"). A limited field test of hyperspectral mine detection was conducted as part of the Airborne Standoff Minefield Detection System (ASTAMIDS) test, with DARPA using the LWIR HS sensor of the University of Hawaii ("AHI").

Polarimetric minefield detection activity has been conducted by Lockheed Martin (Orlando), Raytheon, TRW, and Veridian, with government activity by the British DERA and the U.S. NVESD and Army Waterways Experiment Station (WES). A polarimetric active sensor with an additional thermal IR channel was built for the

Remote Minefield Detection System (REMIDS) program of WES, and again for the ASTAMIDS of NVESD. These sensors were extremely fragile. Another such sensor is now under development for NVESD's Lightweight Airborne Minefield Detection program by Lockheed Martin. In cooperation with the Central Measurement and Signatures Intelligence (MASINT) Office and Air Force Research Laboratory, Veridian and Aerospace have built an LWIR spectral-polarimetric sensor and shown its usefulness in camouflage, concealment, and deception target detection. It has not been tested on mines. Field tests have used active polarimetric and thermal IR for both human-aided minefield recognition (REMIDS) and fully automated detection (ASTAMIDS/Raytheon).

Two types of field tests are considered—data collection (for algorithm development) and performance testing (by an independent agent against requirements, often compounding detection performance with constraints upon processing time or the size of the system). All airborne tests to date have focused on the antiarmor mines (about 12 inches in diameter), and have not yet been including the smaller antipersonnel mine (about 4 inches in diameter). Typical data collection (planning, laying mines or using an existing mined site, integrating available sensor with aircraft, collecting data, truthing it, and post-processing it) can range from $400,000 to $800,000 for each site. Independent performance testing of an established system, involving a greater range of locations, more conditions at each location, and use of an independent test organization, would approximately double the cost to $800,000 to $1.6 million. Collection of minefield data (large orderly arrays of mines placed as representatives of tactical deployment) is perhaps 30 percent more expensive than individual mine detection experiments.

## CURRENT CAPABILITIES AND OPERATING CHARACTERISTICS

Table D.1 summarizes the current field-tested performance of these technologies. The ASTAMIDS tests (index numbers 1, 5, 7, and 8 of the table) were conducted at Fort Huachuca, Ariz., over an arid ground with vegetation cover of sparse grass, low bushes, scrub oak, and cactus, with soil that was a reddish mixture of clay and rocks.

**Table D.1**

**ROC Curve and References for HS/IR Mine Detection Field Tests**

| Mode | # | System | Condition | PD (minefield) | PFA or FAR | Search Rate | Processing Time | Ref. |
|---|---|---|---|---|---|---|---|---|
| Broad-band IR | 1 | ASTAMIDS, Northrop Grumman | 500 minefield encounters, vegetation/rocks | 0.35 | 9.85/sq km | 10.9 sq km/hr | Post-mission (45× real time) | [1] |
| Spectral | 2 | VNIR MS–COBRA ATD (Veridian) | 6-band VNIR spectral sensor, littoral zone surface minefields (white sandy beach) | 0.86 (patterned) 0.94 (scattered) | 0.02 (PFA) (patterned) 0.07 (PFA) (scattered) | 9 sq km/hr | Post-mission (14–44× real time—no requirement) | [2] |
| | 3 | | Littoral/inland surface minefields (rocky beach) | > 0.8 | < 1.5/sq km | 6.5 sq km/hr | Post-mission (14–44× real time—no requirement) | [3] |
| | 4 | Navy MDP and ALRT (Veridian) | Tunable filter polarimetric 3-band VNIR multispectral; ground and air tests | > 0.8 (better than COBRA ATD) | N/A (< COBRA with limited data sets) | N/A | N/A | [11] |
| | 5 | LWIR HS – ASTAMIDS (DARPA) | LWIR HS sensor; buried roadmines (6 mines only) | 0.67 | 471/sq km | 2 sq km/hr | Post-mission | [1] |
| Polari-metric (with LWIR) | 6 | REMIDS (WES, Veridian) | Polarimetric and thermal IR; aids human | 0.989 (patterned) 0.663 (scatterable) | N/A N/A | 1.44 sq km/hr | 24 hr; also achieved PD of 0.92 (patterned) within 2 hr | [4] |

**Table D.1—continued**

| Mode | # | System | Condition | PD (minefield) | PFA or FAR | Search Rate | Processing Time | Ref. |
|---|---|---|---|---|---|---|---|---|
| | 7 | ASTAMIDS, Raytheon | Polarimetric and thermal IR; automated | 0.13 | 2.49 FA/sq km | 10.9 sq km/hr | Post-mission, 2.5× real time | [1] |
| | 8 | | Postprocessed (after test completion)—removed bad data due to sensor registration errors | 0.2 (surface pattern with bad data)<br>0.57 (surface pattern, good data, night only) | N/A | 10.9 sq km/hr | N/A | [5] |

The tests involved over 500 minefield encounters for numbers 1 and 7, while 5 used *extremely* limited data from which no conclusions can be drawn. All tests except 8 were conducted independently, i.e., by an agent other than the developing organization. The COBRA spectral sensor tests (2) were for surface minefields under excellent conditions of white sandy beaches at Eglin Air Force Base, Fla.; the other tests (3) were conducted over more realistic littoral and land regions in Newfoundland, Canada. The REMIDS program (6) used human interpreters to make minefield decisions by inspecting the sensor data.

## KNOWN OR SUSPECTED LIMITATIONS OR RESTRICTIONS ON APPLICABILITY

The best performance for broadband IR is for mines buried under uniform bare soil. Thermal IR performance for surface mines is better at night than at day. It also performs best when the time of observation can be chosen and/or multiple observations can be made at different times of day. Poorest performance occurs with nonuniform soil and soil covered with vegetation (which blocks thermal IR). Overall broadband IR is not as useful as a stand-alone sensor compared with the other sensors considered here. Its performance is limited by the diurnal cycle characteristic and by the high degree of mine-like clutter, with insufficient information available in the broadband IR to allow discrimination. Progress in the study of spectral phenomenology has led to better understanding of the origins of the broadband LWIR signature, as exploited by the Northrop Grumman ASTAMIDS, and now single or multispectral IR bands can be tailored to improve broadband IR performance.

Spectral detection performance is the best of these three sensors but is at the expense of a more complex sensor design. Detailed experiments on the physics of disturbed soil indicate that with a sufficient number of pixels on target, statistically significant discrimination ability between mine target signatures and their local backgrounds occurred within all bands (VNIR, SWIR, MWIR, and LWIR) [6,7]. Joint use of spectral bands has been shown to further improve performance [9,10]. This detector excels in detecting recently buried minefields (less than four weeks old); it is also excellent at surface minefield detection on relatively clear ground (e.g., littoral region [2]). Spectral methods perform most poorly for long-buried mines, over which the soil has returned to its natural state. The performance for buried mines is limited by the fact that buried mines are detected indirectly, through the associated disturbance of surrounding earth, which weathers away. Performance for surface minefield detection is limited by vegetative clutter that covers the mines.

Active polarimetry excels in the detection of surface mines upon uncovered ground, especially when mines are placed in regular patterns. Polarimetry combined with spectral sensing offers the possibility of excellent performance for both surface and buried mines (e.g., spectral/polarimetric sensing). Polarimetry has limited ability to detect buried mines—no specific phenomenology effect has been identified yet. Passive polarimetry is of limited utility because of the wide variation of illumination/receiver geometries and uncertain polarization of the illuminator, yet progress here has been shown.

## ESTIMATED POTENTIAL FOR IMPROVING TECHNOLOGY OVER TWO TO SEVEN YEARS

Performance of IR/HS sensing is more limited by the detection algorithms than by the sensors. Sensor challenges include the engineering of robust, production-quality versions of the research prototypes used in field tests. Finer spatial resolution is believed necessary for accurate mine/clutter discrimination, especially for smaller mines.

For broadband IR, an example of the power of algorithm improvements provides an improvement in detection rate by more than a factor of two [5]. Improvements to pursue for this sensor include: (1) automatic compensation for time of day, thermal heating history, terrain (and confuser) type; (2) use of higher spatial resolution to exploit shape and within-silhouette information (e.g., texture) to aid discrimination against false alarms (FAs); and (3) leverage of minefield detection algorithms to aggregate incomplete mine detections into a more accurate field declaration. While broadband IR is unlikely to meet operational requirements of greater than 0.8 probability of detection (PD) for minefield detection, less than 1.0 FA per square kilometer on its own, an IR sensor may be combined with a complementary sensor.

Spectral limitations include the requirement to use higher-order decision statistics to achieve target/clutter discrimination (e.g., covariance rather than mean). This requires more pixels on target and more spectra needed to set detection thresholds. Spectral sensing also is limited in its ability to discriminate spectral anomalies due to soil disturbance from mine placement from other spectral anomalies, which increases false alarm rates. The understanding of possible false alarms due to burial of other objects has not been explored. Promising evidence of the far-reaching applicability of the disturbed soil phenomenology was obtained by tests that went to six locations around the world, carefully chosen for diversity, and including real minefields in Bosnia and Jordan [7,8,9], yet some additional confirmation is needed. The understanding and ability to model or explain effects of weathering, which gradually erases the effects of mine burial, is limited. Other limits, imposed by perhaps temporary needs to limit sensor complexity (size), include the choice of which multispectral bands to use if full HS capability is not possible, how to adapt these with terrain, and the limits of VNIR spectral sensors to daytime. All these limits, except perhaps solving the weathering effect, can be addressed successfully in a five-to-seven-year program, resulting in a spectral-based mine detection system capable of detecting nearly all recently laid minefields (PD greater than 0.8) at acceptable false alarm rates (less than 1.0 FA per square kilometer). Focus is recommended on reliably designing HS sensors accompanied by adaptive band-subsetting (i.e., adaptively and intelligently combining 100s of bands to 10s of bands for subsequent processing), as opposed to building ever-more-complicated multispectral sensors based on filter wheels or tunable filter cameras.

Polarimetry is most successful for, but limited to, detection of surface mines and minefields. Sensor complexity, in particular achieving multiple pixels on a mine-sized target while achieving the necessary subpixel registration accuracy between the polarization channels, sets the limits on surface minefield detection performance for this sensor.

## OUTLINE OF A SENSIBLE RESEARCH AND DEVELOPMENT PROGRAM

Several points anchor a general philosophy on which a research and development program would be based:

- Harvest existing data that has already been collected to bolster phenomenology understanding and pattern recognition algorithm performance.
- Collect HS data across all bands (VNIR–LWIR), using the HS testing to determine the ultimate band count that is needed and whether these needs can be met with a multispectral sensor or even an agile (adaptively tuned wavelength) sensor. Seek to understand the utility of joint spectral bands that have shown improved performance [10] (e.g., VNIR with SWIR, MWIR and LWIR). Study the effects of aging over several months, and examine both passive and active polarimetry.
- Conduct ground-based tower-mounted data collections to understand phenomenology, and then collect data using a helicopter test bed with a stabilized sensor cavity and the same sensor suite to improve false-alarm and minefield ROCs.
- Do not encumber these data collections with expectations of a performance test, e.g., with sensor size or real-time constraints (but do not ignore their importance for realism—e.g., resolution/aperture).
- Datamine any field test with after-test improvements as exemplified by Radzelovage and Maksymonko [5] to realize full gain from all information obtained.

Table D.2 presents a sequential program based on the above precepts that would be a reasonable, five-to-seven-year, $31 million approach to achieving satisfactory IR/HS minefield detection. It places about a third of the effort on physics and phenomenology, a third on sensor design, and another third on a final performance validation test.

**Table D.2**

**Outline of Five to Seven Year, $31 Million Program Leading to Robust, Satisfactory IR/HS Minefield Detection**

| # | Action | Objective | Cost |
|---|---|---|---|
| 1 | Conduct post-collection analyses on existing data from REMIDS, ASTAMIDS, Hyperspectral Mine Detection Phenomenology, … | Determine domains of success, reasons for failure, and improved processing algorithms | $1,500K (3 programs at $500K each) |
| 2 | Careful analysis of performance of various combinations of 10–20 spectral bands of 50–200 nm width | Understanding of ultimate need for hyperspectral vs. adaptive multispectral vs. fixed multispectral sensor | $1,000K (2 programs at $500K each) |
| 3 | Tower-based collections over VNIR through LWIR bands with HS imaging sensors, with and without active/passive polarimetry | Exploit joint band detection, examine aging effects over several months to determine factors and time constant of erasing burial effects | $1,600K, including cost of multiple spectral instruments |
| 4 | Collection with sensor suite of #3 in helicopter with stabilized sensor cavity, at two diverse U.S. locations over minefields, wide area | Measure false alarm rate and determine/optimize gain afforded by minefields rather than mines only; explore feature clustering minefield algorithms | $3,000K (two locations at $1,500K each, including platform rental and sensors) |
| 5 | Basic algorithm research—begin with probability density functions (PDFs) of hyperspectral features; seek appropriate subsettting methods (e.g., principal components analysis), seek to increase target/clutter distribution separation | Statistical pattern matching/decision theoretic-driven optimization of information across bands; adaptive band selection; data-driven PDFs to ROC curve; aggregation to minefields | $1,500K (2 programs at $750K each) |
| 6 | Spectral data compression | Process to transmit data down limited bandwidth link | $500K |

**Table D.2—continued**

| # | Action | Objective | Cost |
|---|---|---|---|
| 7 | Spatial/spectral integration of information | Determine best joint use of spectral information with high-resolution spatial information (shape, texture) | $1,000K (2 programs at $500K each) |
| 8 | Spectral libraries | Supplementing anomaly detection with matching of signatures of mines or clutter to known spectra to reduce false alarm rate | $1,000K (2 programs at $500K each) |
| 9 | Reliable spectral sensor development | Use above phenomenological and algorithm understanding to design and build minimal-complexity, maximum-performing sensors for handheld, vehicle, and airborne application | $10,000K (3 platforms, leveraging past work at about $2M–$4M each) |
| 10 | Final field test (hand, vehicle, and air platforms) | Demonstrates performance with final sensors | $10,000K (3 demos) |

## REFERENCES

1. G. Maksymonko and N. Le, "A Performance Comparison of Standoff Minefield Detection Algorithms Using Thermal IR Image Data," in *Detection and Remediation Technologies for Mines and Minelike Targets IV,* A. C. Dubey, J. F. Harvey, and J. Broach, eds., Seattle: International Society for Optical Engineering, April 1999, pp. 852–863.

2. R. R. Muise et al., "Coastal Mine Detection Using the COBRA Multispectral Sensor," in *Detection and Remediation Technologies for Mines and Minelike Targets,* A. C. Dubey, R. L. Barnard, C. J. Lowe, and J. E. McFee, eds., Seattle: International Society for Optical Engineering, 1996, pp. 15–24.

3. K. S. Davis et al., "Coastal Battlefield Reconnaissance and Analysis (COBRA) System Participation in the Joint Countermine (JCM) Advanced Concept Technology Demonstration (ACTD) Demo II at MARCOT 98 Quick Look Report," NSWC Dahlgren, CSS, September 1998.

4. S. A. Pranger et al., "Standoff Minefield Detection System, Advanced Technology Transition Demonstration (STAMIDS ATTD): Program Summary," Technical Report EL-92-10, U.S. Army Engineer Waterways Experiment Station, Vicksburg, Miss., 1992.

5. W. C. Radzelovage and G. B. Maksymonko, "Lessons Learned from a Multi-Mode Infrared Airborne Minefield Detection System: What Worked and What Did Not for Automatic Surface and Buried Minefield Detection," in *Proceedings Infrared Information Symposia, Third NATO IRIS Joint Symposium,* Vol. 43, No. 3, July 1999.

6. A. C. Kenton et al., "Detection of Land Mines with Hyperspectral Data," in *Detection and Remediation Technologies for Mines and Minelike Targets IV,* A. C. Dubey, J. F. Harvey, J. Broach, and R. E. Dugan, eds., Seattle: International Society for Optical Engineering, April 1999, pp. 917–928.

7. A. Smith et al., "Hyperspectral Mine Detection Phenomenology Program: Mine Spectral Collections and Data Archive," Report to

U.S. Army CECOM, Fort Belvoir, Va., ERIM International Report 10012200-15-T, contract DAAB07-98-C-6009, March 1999.

8. C. R. Schwartz et al., "Hyperspectral Mine Detection Phenomenology Program: Spectral Mine Detection," Report to U.S. Army CECOM, Fort Belvoir, Va., ERIM International Report 10012200-17-F, contract DAAB07-98-C-6009, March 1999.

9. W. Malila et al., "Hyperspectral Mine Detection Phenomenology Program: Spectral Mine Detection Study (Phase II), Initial Joint Spectral Region Mine Discrimination Performance Assessment," Report to U.S. Army CECOM, Fort Belvoir, Va., ERIM International Report 10012200-30-F, contract DAAB07-98-C-6009, March 2000.

10. A. C. Kenton et al., "Joint Spectral Region Buried Land Mine Discrimination Performance," in *Detection and Remediation Technologies for Mines and Minelike Targets V*, A. C. Dubey, J. F. Harvey, J. Broach, and R. E. Dugan, eds., Seattle: International Society for Optical Engineering, 2000, pp. 210–219.

11. J. S. Taylor et al., "Laboratory Characterization and Field Testing of the Tunable Multispectral Camera," in *Detection and Remediation Technologies for Mines and Minelike Targets VI*, A. C. Dubey, J. T. Broach, and V. George, eds., Seattle: International Society for Optical Engineering, 2001, pp. 1247–1258.

Appendix E

# GROUND-PENETRATING RADAR (PAPER I)

*Lawrence Carin, Duke University*

This appendix focuses on the application of ground-penetrating radar (GPR) to landmine detection. We address the particular application for which the sensor is near the ground.[1]

## BASIC PRINCIPLES AND MINE FEATURES

GPR senses electrical inhomogeneities, with these manifested, for example, by the electrical contrast of a metal mine in the presence of a far-less-conducting soil background. Of much more difficulty, GPR senses the electrical inhomogeneity caused by a dielectric (plastic) landmine in the presence of soil. Often this contrast is very weak, implying that the landmine GPR signal is very small. This is exacerbated by the fact that there are many electrical contrasts that may exist in the landmine problem, which significantly complicate sensing. For example, the largest contrast typically exists between the air and the soil, and therefore GPR is typically characterized by a very large "ground bounce." If a landmine is buried at a shallow depth, such that the available bandwidth implies limited resolution, the often weak landmine signature will be "buried" in the very strong ground-bounce return. This implies that bandwidth (resolution) plays an important role in defining the target depths at which a landmine may be observed by GPR. We also note that natural subsurface inhomogeneities, such as rocks, roots, surface roughness, and soil heterogeneity (e.g., pockets of wet soil), also yield a signature

[1] I do not consider the use of radar for airborne, wide-area sensing because it was indicated to be of less interest to RAND and the Office of Science and Technology Policy.

to a GPR sensor; such clutter represents the principal source of false alarms.

## CURRENT STATE OF DEVELOPMENT

GPR is one of the oldest landmine technologies, probably second only to induction sensors. Nevertheless, development is still at a relatively early stage, as only very recently have electromagnetic models been developed to aid in phenomenological understanding. Consequently, only recently has enhanced understanding of the underlying phenomenology been exploited in new systems, such that the ultimate potential of GPR can be realized.

As for recent efforts, GPR is being employed by Cyterra in the context of the U.S. Army Handheld Standoff Mine Detection System program. The Cyterra system is based on multiple spiral antennas and operates in the frequency domain. Another important system has been developed in Germany by Gunter Wichmann, with this technology now being pursued in the United States by NIITEK. The so-called Wichmann system operates in the time domain, with a very large bandwidth (from approximately 200 MHz up to 10 GHz). A novel antenna design has significantly reduced the antenna "self clutter," resulting in a very-high-resolution transient waveform. This system is not at the stage of development of the Cyterra system, but it represents a significant enhancement in GPR technology (the result of three decades of development in Germany).

With regard to development of modeling tools to understand GPR phenomenology, there are several universities that have directed significant attention on GPR for landmine detection, including Duke, Georgia Tech, Northeastern, and Ohio State. The state of model development is now becoming quite sophisticated; it is now possible to rigorously model a three-dimensional GPR system on a computer.

## CURRENT CAPABILITIES AND OPERATING CHARACTERISTICS

It is difficult to separate GPR performance from the particular system and classification algorithms in question. Moreover, the Cyterra system combines both GPR and induction, and therefore the performance of this system is generally not simply based on the GPR characteristics.

It can be said, however, that significant strides have been made in the last several years in the context of GPR sensors. There is a "blind grid" of buried landmines and clutter developed by the Joint Unexploded Ordnance Coordination Office at Fort A. P. Hill, Va. The investigator is told where to sense (on the grid), but the subsurface target identity is unknown. Using the Wichmann system, with very preliminary algorithm development, approximate results are a probability of detection of 0.8 at a probability of false alarm of 0.1. It is anticipated that these results will improve as the algorithms are enhanced. It is believed that the Cyterra system achieved comparable results when it deployed its GPR alone (although the results were far better when the Cyterra GPR and induction sensors were fused). This test site was designed to be particularly challenging.

It is important to emphasize that the quality of the signal-processing algorithms plays a very important role in the ultimate effectiveness of a GPR system. For example, GPR has been used in the context of the Ground Standoff Mine Detection System (GSTAMIDS) program (vehicle program). Previously, GPR produced an unacceptably high number of false alarms for on-road sensing. The recent development of such algorithms as hidden Markov models has significantly aided GPR classification performance within the GSTAMIDS program, and these algorithms (and performance) continue to improve.

## KNOWN OR SUSPECTED LIMITATIONS ON APPLICABILITY

As indicated in the first section, GPR is not a landmine sensor—it is an electrical (and possibly magnetic) contrast sensor. Any such contrast will register a signal to GPR, and often subsurface clutter can manifest a signal comparable to that of a landmine. Moreover, in the context of plastic or dielectric (wood) mines, the properties of the soil play an integral role in ultimate GPR utility. For the same mine, a given GPR can be effective or ineffective depending on the soil properties. This implies that such elements as rain, or the lack of rain, can play a critical role in the context of the target-soil contrast, affecting ultimate sensor performance.

There are also important issues in the context of very small plastic mines buried at shallow depths (e.g., flush-buried). The strong ground-bounce return can "mask" the much smaller signal from the landmine. This is why bandwidth is a particularly important issue: A higher-bandwidth GPR system will have enhanced resolution to deal with shallow-buried targets.

## ESTIMATED POTENTIAL FOR IMPROVEMENT (OVER TWO TO SEVEN YEARS)

GPR has limitations, as indicated in the previous section. However, it is likely to be an integral component of any landmine sensor suite because a properly designed system, with appropriate signal-processing algorithms, has demonstrated significant potential in quickly sensing landmines. In concert with other sensors, such as induction and nuclear quadrupole resonance (NQR), GPR constitutes a very powerful technology.

There is significant potential for GPR improvement in the next two to seven years. As indicated in the Current State of Development above, only recently has the fidelity of electromagnetic models put developers in a position where they can *understand* the various tradeoffs in GPR design, such as antenna configuration, bandwidth, and polarization. Now that this insight can be transitioned to the development of improved GPR sensors, there is significant potential to realize significant improvements in GPR performance, particularly in the context of a multisensor suite.

Issues that should be addressed include examination of the effects of antenna design on ultimate performance. The Cyterra and Wichmann systems, for example, are based on entirely different antenna systems, although both are yielding promising results. It is of interest to carefully examine the relative strengths and weaknesses of each system (and others) such that an overall improved design can be realized. Moreover, within the context of a multisensor suite, one should examine design of the GPR such that it is most effective, for example, as a prescreener for such sensors as NQR (which appears to be a promising confirmation sensor).

## OUTLINE OF A SENSIBLE RESEARCH AND DEVELOPMENT PROGRAM

The ultimate multisensor platform is almost surely going to include GPR as an integral component. I would therefore recommend first performing a systems-level study to identify the role of GPR and how it will be manifested in the context of multiple sensors (e.g., GPR and induction may be prescreeners for NQR). Once these systems-level issues are addressed, it is desirable to design a GPR system that would meet the objectives of the overarching systems-level mission. The ultimate system construct may constrain or motivate such items as GPR bandwidth, polarization, antenna design, and size.

In addition, there are several existing systems (e.g., the Cyterra and Wichmann systems) that operate in very different ways, although each has yielded encouraging results. It is of significant interest to gain a further understanding of these systems, based on measurements and modeling, such that one may take this insight to develop the next-generation GPR system. This research should build on and extend recent encouraging research in the area of modeling and radar development.

It should also be emphasized that, as indicated above, within the context of GPR, one cannot separate the system from the associated signal processing and ultimately how the system will be deployed by the user. It is strongly recommended that the modeling, sensor design, and GPR signal processing be strongly coordinated from the outset. Typically, GPR systems are designed prior to considering how the data will be processed, and therefore processors must deal with the subsequent GPR data without the ability to motivate change in underlying sensor design. GPR must be viewed in the context of an overarching system, accounting for the other sensors that will be deployed as well as the ultimate manner in which the data will be processed and presented to the user.

## REFERENCES

The best source of reference to field-testing papers is likely to come from reports delivered to the sponsors (Countermine Office, Fort

Belvoir, Va.) and from the proceedings of the International Society for Optical Engineering's AeroSense Conference (primarily from 1998 to 2002).

Appendix F

# GROUND-PENETRATING RADAR (PAPER II)

*James Ralston, Anne Andrews, Frank Rotondo, and Michael Tuley, Institute for Defense Analyses*[1]

## BACKGROUND AND SUMMARY

It is understood that information on the performance of individual landmine sensors and groups of sensors is being considered in the context of a mine remediation process that includes site identification, the location of individual mines, and appropriate mine-by-mine remediation measures. This understanding must include several key points, detailed in the following paragraphs:

**Study the remediation process.** This must be done before attempting to model the process. It must also be recognized that remediation includes any measures applied to the mined area before sweeping with detection systems. For example, it is sometimes the practice to use heavy rollers in an attempt to detonate some mines before sweeping. This can be effective in preemptively disposing of many mines at relatively low cost and risk, but the aftermath can be detrimental to the function of detecting sensors if some unexploded mines are inadvertently buried to greater depths, misoriented, or tilted so that they become hazardous to probe manually, or if their soil context is changed in ways that may affect certain sensors, such as radar, acoustics, or infrared. Any attempt to draw conclusions about enhanced remediation on the basis of modeling must first ascertain that the model's assumptions are realistic.

[1]Originally published by the Institute for Defense Analyses, Alexandria, Va. Reprinted by permission.

**Understand local differences.** For a variety of political, cultural, and economic reasons the process of landmine remediation proceeds differently in different countries. This must be respected in assigning overall costs to various stages of the process. For example, operations that would be "prohibitively labor intensive" in one country may be acceptable in nations that have large populations of unemployed farmers. Elapsed time or man-hours is not generally a good proxy for cost. Properly administered, mine remediation in such countries may represent economic as well as humanitarian assistance.

**Fusion is critical.** The goal should be to find not the best sensor but the best combination of sensors. Fusing the two best individual sensors does not necessarily lead to the best fused performance. Data collections should emphasize establishing statistical correlations between detections and false alarms of all sensors, rather than simply measuring individual sensor performance.

**Performance is situational.** All sensors will work better in some environments and against some threats than in and against others. The best choice of sensors and fusion techniques will depend on the specifics of each environment.

**Prospects for improvement.** The most likely research and investment areas leading to possible significant reductions in mine remediation rate are as follows:

- Reducing false alarms by emphasizing fusion of multiple sensors or the use of multiple stages of detection and confirmation sensors.
- Identifying the specific classes of false alarms whose characteristic features can be recognized and confidently rejected. The classes may include particular man-made objects (e.g., soft-drink cans) as well as naturally occurring soil phenomena. The feature set investigated should not be limited to the output of a single sensor type.
- Seeking cost reductions in the most advantageous suite of sensors and remediation techniques, once it is identified.

Andrews, Ralston, and Tuley [1] document the results of an Institute for Defense Analyses (IDA) assessment of the state of current research on ground-penetrating radar (GPR) as applied to countermine and unexploded ordnance (UXO) clearance. Despite significant long-term investment in GPR for mine and UXO detection, it remains true that no GPR system that meets operational requirements has yet been fielded; however, recent advances in several mine detection radars under development have produced significant improvements in detection performance and false-alarm mitigation over what was achievable only a few years ago. The authors' report examines existing GPR research and development efforts with emphasis on missions where GPR has the potential to provide a unique capability and to achieve operationally meaningful performance. It identifies data collections and analyses that will be necessary both to make decisions about the suitability of GPR for particular missions and to achieve performance gains necessary for operational utility.

The Joint Unexploded Ordnance Coordination Office (JUXOCO) sponsored a GPR workshop held at IDA in June 1999. Investigators from all currently funded GPR efforts in countermine and UXO were invited to present their work. An independent panel representing government, federally funded research and development center, and university expertise in GPR was assembled to assist the government in the assessment role.[2] Panel discussions were held during the course of the three-day workshop and a one-day follow-on meeting. The report of Andrews, Ralston, and Tuley [1] does not an attempt to express a consensus of the panel, which likely does not exist; however, the comments and insights provided by panel members are reflected in the emphasis and conclusions of this report.

## CONCLUSIONS

The principal conclusions of *Research on Ground Penetrating Radar for Detection of Mines and Unexploded Ordnance* [1] are as follows:

- Phenomenology controlling performance is not sufficiently well understood. Advanced development work must be preceded by concomitant research understanding.

---

[2]Presentations to the workshop are summarized in Andrews, Ralston, and Tuley [1].

- Too little analysis has been carried out on the data that have been obtained. A synergistic analysis effort that stretches across programs might provide real dividends.
- While there are exceptions, current system performance is typically limited by false alarms. That is, detection is clutter limited, not noise limited. Only when target and clutter characteristics are both well understood can signal processing be applied effectively.
- Much more effort has been spent studying target characteristics than has been spent on clutter. Efforts defining target signatures are necessary, and target-related research should continue; however, substantial efforts must be focused on clutter research and data collection.
- Predicting performance requires understanding sensitivities to the environment. Models will not provide the realistic data useful in algorithm development until the understanding of clutter is improved.
- Incorporation of diverse expertise in sensor hardware, algorithm development, modeling, and testing has been beneficial. The Multidisciplinary University Research Initiative (MURI) and the red team approach to the Handheld Standoff Mine Detection System (HSTAMIDS) program have resulted in a better understanding of the sensor functionality and performance improvements.
- There is a need for controlled, repeatable testing to evaluate sensor performance independent of operator skill and technique, and not subject to uncontrollable alterations in the environment. This capability is important for comparing different sensors and tracking changes in performance with sensor modifications.

## RECOMMENDATIONS

1. GPR countermine performance is limited by clutter, and clutter is not well understood.[3] Thus, the focus of research should be on

---

[3]*Clutter* is defined to be returns identified by the sensor system as targets that do not correspond either to intended targets or to random system noise; that is, clutter repre-

defining, understanding, and measuring clutter. To that end, the following steps should be undertaken:

— Determine the range of clutter and target data needed to support system design decisions, algorithm development, and modeling research.

— Build a suite of research-quality data-collection instruments not constrained by operational requirements.

— Collect and analyze clutter and target data, with a focus on clutter. Data collection should be driven by three concerns: better understanding clutter characteristics, providing training and test data for signal-processing algorithm development, and providing both input and validation data for electromagnetic model development.

— Table F.1 provides our recommendations for the system design and parameter space to be covered by the instruments and the data collection. These recommendations are for reasonable, notional parameters for the instruments and the experiments, but they do not represent the results of a rigorous study of the trade space or practical engineering considerations. As such, final designs should be based on an extensive red team effort involving hardware engineers, signal processors, modelers, and test designers.

— Develop a research program to provide the necessary knowledge of clutter characteristics. Such a program should involve a careful physical and electromagnetic description of environments of interest, ranked in order of importance. These could be used to prioritize data collections. Clutter is highly variable, and that complicates its description. The focus of the research should be an attempt to group clutter into a limited number of classes relevant to system design. To that end, a careful evaluation of a combination of statistical and discrete approaches for clutter characterization is warranted.

---

sents real sensor responses to discrete items or environmental conditions that are not of interest.

**Table F.1**

**Data Collection Matrix**

| Parameter | Value Range |
|---|---|
| Frequency range | 200 MHz–6 GHz |
| Polarization | Full |
| Grazing angle | Full hemisphere |
| Aspect angle | Full hemisphere |
| Road/terrain/area | Increasingly complex media, small patches |
| Target type/configuration/ quantity | Individual target interrogation:<br>—Buried mines<br>—UXO<br>—Discrete clutter objects<br>—Standard metal and dielectric targets |
| Spatial resolution | Less than minimum target size, best attainable with radar, centimeters |
| Waveform | Stepped frequency |
| Azimuthal processing | Three-dimensional SAR |
| Antenna height | Close coupled to earth |

— Support research on the characterization of electromagnetic propagation and scattering in soils. Investigate a statistical paradigm similar to the atmospheric weak-scattering case. Bolster theoretical analysis with carefully calibrated measurements and computer modeling. Efforts should begin on simple, well-characterized media. As understanding is gained, more complex compositions should be tackled.

2. We should do a better job of exploiting data from current programs. There are two important facets of such an effort:

— Make data and specific analyses deliverable from contractors. Every effort should be made to ensure that data collections and analyses serve the broader goals of the countermine program.

— Set aside resources for independent analysis of data. Such efforts provide potentially valuable insights that are not likely to come out of program-driven analyses. An example is the red team analysis of HSTAMIDS[4] data, which provided significant input to focus system improvements.

3. The HSTAMIDS red team is an example of how accessing a larger body of knowledge in the countermine area can pay dividends for a specific program. Research results coming out of MURI and applied to data from the BoomSAR, Wichmann, and Geo-Centers systems show significant performance improvements. Such interactions should be encouraged through a red team approach to system engineering decisions.

4. Measurement, modeling, and detection/discrimination algorithm development must be tightly integrated. As discrimination of mines from clutter is typically the problem faced by mine detection systems, discrimination algorithm development is the key to performance improvement. Algorithm success depends on the signals provided.

5. Sensors delivered to the government at the end of programs should be well documented and well calibrated.

6. Existing platforms from other Department of Defense programs should be leveraged to the extent possible. Specifically, the Defense Advanced Research Projects Agency ultra-high-frequency ultra-wide-band synthetic aperture radar (DARPA UHF UWB SAR) and the Army Communications–Electronics Command tactical unmanned aerial vehicle SAR should be tasked for data collection and baseline performance determination for countermine and UXO detection.

7. Develop protocols and equipment for standardized sensor testing.

8. Other specific recommendations are summarized in Table F.2.

---

[4]HSTAMIDS is under development for the U.S. Army. The system incorporates both GPR and electromagnetic induction (EMI) sensors.

**Table F.2**

**Other Recommendations**

| | |
|---|---|
| Soil characterization | Develop a statistical description of soils, patterned on atmospheric physics<br>Develop numerical modeling approaches that accurately represent realistic soils<br>Initiate a measurement program to support above |
| Discrimination | Continue modeling efforts to identify discriminants<br>Use above data collection for signal processing<br>Investigate utility of polarization<br>Investigate utility of spectral response<br>Curtail complex natural resonance research<br>Curtail 3rd harmonic research |
| Fusion | Require analysis and reporting of target and clutter statistics for current data<br>Make raw and processed data deliverable.<br>Initiate independent analysis<br>Task collection of coregistered data sets for forward-looking radar with NQR and forward-looking and down-looking radar |

## HANDHELD DETECTOR PERFORMANCE

Below we survey results of recent tests of HSTAMIDS. This sensor system incorporates both a GPR and a metal detector. In the most recent testing [2], both sensors were used, and test results reflect a fusion of both sensor indications. In earlier tests [3], the performance of the individual sensors was separately tabulated. Here we use HSTAMIDS as a proxy for a state-of-the-art GPR sensor for small mine detection to quantify what performance improvements are necessary.

All of these results are influenced by the location and procedures of the particular test, as well as previous engineering design choices in the systems tested. Thus they are not easily extrapolated to general statements about capability. This points to important considerations about testing and evaluating detection systems. That is, there is a need to evaluate sensors in a standard environment on a common and reproducible target set where the influences of the operator are eliminated. This is particularly important for tracking system development and for evaluating competing sensor concepts. In field tests, true performance of sensors can be masked by contributions from footprint and coverage, as well as operator skill or fatigue, exploitation of visual cues, familiarity with the test site, or the means of presenting the data to the operator.

The HSTAMIDS Operational Requirements Document sets the requirement for detection of surface and buried antipersonnel and antitank mines at probability of detection (PD) = 0.90, with a false alarm rate not to exceed 0.6 per square meter [4]. Historically, handheld GPR sensors have been stressed by low-metal-content antipersonnel mines. (See, for example, Andrews et al. [5], where systems incorporating GPR and EMI sensors were tested in 1996.) Recent modifications in sensors under the HSTAMIDS program have resulted in performance improvements over what was achieved only a few years ago [2].

## CLUTTER

A review of baseline data quickly leads to the conclusion that, in many cases, the fundamental problem of GPR performance is not the absence of a sufficient mine- or UXO-generated signal for the radar to detect.[5] Rather, the problem is the multitude of signals originating from surface and buried clutter. Here, the concern is separating target signals from clutter signals. Thus a thorough understanding of clutter becomes fundamental to understanding GPR performance and limitations.

We define *clutter* as returns identified by the sensor system as targets that do not correspond to intended targets or system noise, that is, real sensor responses to discrete items or environmental conditions that are not of interest. Clutter might be considered as a set of area- or volume-extensive attributes of the environment in which a GPR must work. Conversely, we might think of clutter as a collection of discrete, but undesired, targets to be separated from those we desire to detect. It is likely worthwhile to employ a mix of both views of

---

[5]An exception is likely to be detection of dielectric mines buried in soils with similar dielectric constants, where little or no contrast may exist.

clutter, and we do so here by defining two clutter study modalities: volume and discrete.

In either case, clutter statistics, which will quantify the ability of a feature or set of features to separate targets and clutter, must be measured for each potential discrimination feature of interest and will differ for each radar configuration. A library of target and clutter measurements taken in a variety of environments can be used to determine the robustness of clutter suppression approaches.

Features of discrete clutter objects (e.g., rocks, roots, cans, water-filled inclusions) may allow for discrimination and identification to reduce false alarms. It is possible but not currently known whether a significant fraction of false alarms arise from a small number of discrete types of clutter. This makes identification of the sources of false alarms an imperative part of any sensor improvement clutter study. If discrete objects or features of the ground can be identified and characterized, they can be screened out where they differ sufficiently from targets.

Studies of clutter are often neglected because of the desire to obtain data that are universally useful. Because the sensor must perform in a highly variable and continuously changing clutter environment, however, ways must be found to understand clutter. Framing experiments to study clutter is difficult because any study will be of the specific clutter at a specific site as seen by a specific instrument. There will be great variability in the clutter itself, based on uncontrollable variables such as geology, climate, and history of use. Clutter at the same site may have temporal variations depending on recent weather patterns. Further, the clutter will depend on features of the radar itself, such as grazing angle, spot size, resolution, frequency band, and polarization, as well as the processing. A clutter experiment therefore will require careful research planning, data collection, and analysis. Ongoing programs provide numerous data collection opportunities. There is a need to make sure the right data are taken, that they are analyzed, and that the analysis parameterizes the data in a meaningful way.

Recently, DARPA conducted a data-collection program that focused on understanding clutter for the buried mine and UXO problem [6]. Numerous sensors were tasked to survey four 1-hectare sites, with the goal of producing co-registered clutter maps from multiple sensor modalities for magnetometry, EMI, radar, and infrared sensors. All three radars in the study were sensors of opportunity, so in any event the data collected explored dimensions determined by previous design choices. The experiment also experienced navigation difficulties that made interpretation of the data difficult. Nevertheless, algorithm work done by Paul Gader on Geo-Centers radar data set resulted in a many-fold decrease in the density of false alarms at a comparable PD [7].

The necessary clutter study will require an effort such as this. This experiment should be conducted in a variety of clutter environments on well-characterized sites. Research-grade sensors with the flexibility to explore the widest accessible parameter space are required. Only through an effort such as this can we build the library of clutter data necessary for making engineering design choices; supporting modeling of real-world conditions, signal processing, and algorithm development; and for determining the robustness of sensor performance.

## GPR SYSTEM CONSIDERATIONS

GPR design is complex and challenging because of the array of hardware and system choices and the coupling of many of those choices. This section provides a brief discussion of some of the choices that can be made and their implications.

The most fundamental choice in GPR is the center frequency and bandwidth of the radar. Low frequencies provide improved soil penetration; the depth at which targets must be detected and the soil types within which they must be detected drive the choice for the lowest frequencies to be transmitted. For example, UXO detection would generally call for lower frequencies than mine detection because of the greater depths at which targets may be located. Practical limits on low-frequency performance are often determined by the maximum size of the antenna that can be deployed. Range resolution is governed by bandwidth, with the achievable resolution given as the speed of light in the medium divided by twice the bandwidth:

$$\Delta R = \frac{c}{2B\sqrt{\mu_r \varepsilon_r}},$$

where $c$ is the speed of light, $B$ is the bandwidth, $\mu_r$ is the relative permeability, and $\varepsilon_r$ is the relative dielectric constant. Thus, if high resolution in range is desired, wide bandwidth is required, and the higher the center frequency, the narrower the percentage bandwidth for a given resolution and the more straightforward the radar design job. Because of the dispersive properties of soil, high frequencies will be attenuated more than low frequencies. Rather than considering the waveform that is transmitted, the GPR designer must plan his processing and detection strategies around the expected spectrum of the return after propagation to the target, reflection, and propagation back to the radar antenna. Thus, having low frequencies that penetrate well may be of little consequence if the detection algorithm depends on fine resolution and the higher frequencies that provide bandwidth are severely attenuated. The chosen frequency regime also controls less obvious radar characteristics, such as achievable cross-range resolution, in SAR systems and the level of radio frequency interference (RFI) with which the system must contend.

Most GPRs for mine detection are wideband devices because good range resolution is required to separate targets from clutter. Two general approaches to obtaining wideband performance are available to the system designer. Each has advantages and disadvantages. The first utilizes waveforms having time-bandwidth product that is near unity. These systems are represented by the family of impulse radars that have been developed for ground-penetration missions. The major advantages of an impulse radar are that lower dynamic range receivers are required to discriminate against clutter, the waveform generation time is short, and a high-range resolution display is available with little or no processing. The major disadvantages are the need to control radio frequency dispersion over a wide instantaneous bandwidth, susceptibility to RFI because of the wideband receiver front end, the need for very-high-speed analog-to-digital converters (or the inefficiency of a sampling oscilloscope approach) for waveform capture, and difficulty in controlling details of the transmitted spectrum.

The alternative to impulse is to employ a waveform with a time-bandwidth product much greater than one. Such systems have been implemented using stepped frequency, linear FM (frequency modulation) chirp, or phase codes. The major advantage of stepped frequency or linear FM chirp is that the frequency spectrum can easily be chosen to fit what the designer considers optimum. In fact, notches can even be placed in the transmitted spectrum to avoid interference with or by other systems. Stepped-frequency waveforms in particular allow narrow instantaneous receiver bandwidth, lower bandwidth analog-to-digital converters, and wider dynamic ranges. This last advantage is often offset by a need for the wider dynamic range because the large surface clutter return and target returns are not temporally separated as they are in an impulse system. Other advantages of high time-bandwidth product waveforms are higher average powers and an ability to tailor the frequency response on receive through processing. Phase and amplitude calibration and equalization are easily accomplished at each discrete frequency step. The major disadvantages are the required dynamic range mentioned above and the time required to generate one complete waveform.

A waveform and bandwidth having been chosen, the GPR designer must implement an antenna commensurate with the bandwidth. Antenna design becomes particularly critical in systems whose geometry provides little standoff from the surface of earth. A major obstacle in using wideband antennas is eliminating internal reflections over their entire frequency band. Such internal reflections result in antenna "ringing" that can hide target returns in systems that operate close to the surface. In down-looking systems, that problem is exacerbated by the very large surface clutter return that may also reverberate within a poorly matched antenna structure. Closely coupled antennas can reduce that problem, as can the use of cross-polarized antennas that tend to discriminate against the surface clutter return. While no designer would intentionally choose an antenna known to produce significant internal reflections, two design approaches are generally viable. In one, the designer makes heroic attempts to reduce antenna internal reflections, thereby simplifying the signal-processing problem. Such an approach is illustrated by the Wichmann radar, where the array antenna is designed to minimize both internal reflections and reverberation between the antenna face and the ground surface. The second option is to have a certain amount of antenna internal reflections and take those out in signal processing. This is most easily done with a stepped-frequency system, where the internal reflections can, in principle, be measured at each frequency and then coherently subtracted from the return. The flaw in such an approach is that the reflections will depend to some extent on the details of the surface clutter return, and as that changes, coherent subtraction may be less effective. Internal reflections are of less concern in standoff radars, but at the low frequencies often employed in GPR they may be a problem even in that case. In particular, at low frequencies the entire structure on which the antenna is mounted becomes part of the radiating structure, and reverberations may linger in time. That problem has been noted in several GPR implementations.

Signal-processing and display options are strongly driven by the waveform choice and the antenna implementation. For example, with a single antenna that is manually scanned, it is very difficult to generate a display output more sophisticated than a simple one-dimensional range profile. Such a display is available with little or no processing from an impulse system and with simple pulse-compression processing from a large time-bandwidth product system. Linear array antennas or those antennas that scan across mine lanes provide more flexibility in display. A linear array can be used to provide a waterfall plot of time (range) images closely spaced in the cross-track direction, as in the Wichmann radar; a real aperture two-dimensional image, as in the Geo-Centers display; or a form of synthetic aperture image. A scanned antenna can also be used to produce a synthetic aperture image. Finally, scanning in both cross-track and down-track allows formulation of three-dimensional images. Doing so, however, requires careful attention to knowledge of antenna position and correction of propagation effects within the soil.

There should be a clear connection between an operational concept for GPR in countermine operations and a focused plan for conducting the necessary supporting research and obtaining needed engineering design data. The focus of the JUXOCO report [1] is on defining the research necessary to design an optimal operational system, not to design the operational system forthwith. As such, it is important to remember that the recommendations herein are for instrumentation and experiments to collect necessary data rather than to support operational requirements. We will generally want data-collection experiments to cover a wider range of operating parameters than would be practical for a fielded system so that we can be confident that the limits of the operating system are optimally chosen.

## REFERENCES


1. A. Andrews, J. Ralston, and M. Tuley, *Research on Ground Penetrating Radar for Detection of Mines and Unexploded Ordnance: Current Status and Research Strategy,* Alexandria, Va.: Institute for Defense Analyses, D-2416, December 1999.
2. F. Rotondo, E. Ayers, A. Calhoun, E. Rosen, and L. Zheng, *Engineering Development Test Results for the Handheld Standoff Mine Detection System: May 2000,* Alexandria, Va.: Institute for Defense Analyses, D-2510, March 2000.
3. F. Rotondo, E. Rosen, and E. Ayers, *Test Methodology and Results for the Handheld Standoff Mine Detection System in Check Tests 1 and 2: June–October 1999,* Alexandria, Va.: Institute for Defense Analyses, D-2443, March 2000.
4. Operational Requirements Document for the Handheld Standoff Minefield Detection System, U.S. Army Training and Doctrine Command, August 19, 1995.
5. A. M. Andrews, T. W. Altshuler, E. M. Rosen, and L. J. Porter, *Performance in December 1996 Hand-Held Landmine Detection Tests at APG, Coleman Research Corp. (CRC), GDE Systems, Inc. (GDE), and AN/PSS-12,* Alexandria, Va.: Institute for Defense Analyses, D-2126, March 1998.
6. V. George and T. W. Altshuler, "Summary of the DARPA Background Clutter Experiment," Proceedings EUROEM Conference 1998, Tel Aviv, Israel; George, V. and Altshuler, T. W., "Summary of the DARPA Background Clutter Experiment," Proceedings FUZZ-IEEE 1998.
7. P. Gader, J. Keller, and H. Liu, "Landmine Detection Using Fuzzy Clustering in DARPA Backgrounds Data Collected with the Geo-Centers Ground-Penetrating Radar," in *Detection and Remediation Technologies for Mines and Minelike Targets III,* A. C. Dubey,

J. F. Harvey, and J. Broach, eds., Seattle: International Society for Optical Engineering, 1998, p. 1139.

Appendix G

# ACOUSTIC/SEISMIC METHODS (PAPER I)

*James Sabatier, University of Mississippi*[1]

## BASIC PHYSICAL PRINCIPLES AND MINE FEATURES EXPLOITED BY THE TECHNOLOGY

Acoustic-to-seismic (A/S) coupling–based mine detection is based on the ability of sound to penetrate the ground and excite resonances in buried compliant objects [1]. Sound produced in the air efficiently couples into the first 0.5 m of the soil because of the porous nature of weathered ground resulting in acoustic vibrations that are sensitive to the presence of buried mines. This phenomenon has been termed "acoustic-to-seismic coupling" in the research literature. Off-the-shelf loudspeakers are used as the sound source and readily available laser Doppler vibrometers (LDVs) are used to measure the increased ground vibrations due to the presence of the buried mine [2]. The use of airborne acoustics for mine detection exploits three new phenomena that previously have not been explored for the purpose of buried mine detection. First, a landmine is a man-made, acoustically compliant object that is much more compliant than soils. This results in a high-vibration contrast between soils and buried mines that does not occur because of the presence of rocks, roots, and other man-made solid objects, such as concrete and metal. Second, the mine is a nonporous object that offers additional contrast to the porous soil in the presence of coupled sound. Third, the interface between a mine and the soil has been shown to be nonlinear; the interface between the soil and the

[1]Originally published (in another format) for the U.S. Army, January 2003.

mine is not continuous when vibrating. This phenomenon results from the strong acoustic compliance of the mine compared with soil. This nonlinear phenomenon allows for unique measurements that are in theory absolutely free of false alarms [3].

## STATE OF DEVELOPMENT

*Is the technology in basic laboratory research, or has it been field tested? What organizations and research programs are examining it? What are the realized or projected costs of field testing?*

Extensive laboratory and field research of A/S coupled buried mine detection has been performed in recent years. Because of the applied research development level, an operational configuration for system-level development test and evaluation and operational test and evaluation has not been established. Both nonlinear and linear acoustic techniques are currently being field tested by the University of Mississippi's National Center for Physical Acoustics (NCPA) and Stevens Institute of Technology (SIT). NCPA, working with Planning Systems Inc. of Slidell, La., and MetroLaser Inc. of Irvine, Calif., is developing an acoustic-based antitank mine detection data collection platform with multiple laser vibrometers for the U.S. Army Communications–Electronics Command Night Vision and Electronic Sensors Directorate (NVESD), Fort Belvoir, Va., that will move at a few kilometers per hour with a 1-m-wide swath. The Army Research Office and the Office of Naval Research have also provided additional funds for related research to NCPA. Current budgets for the NCPA and SIT efforts are provided by the congressional plus-ups and have totaled approximately $3 million per year. Because of the expressed interest of the sponsor, the majority of the funds are used for the antitank mine detection effort.

## CURRENT CAPABILITIES AND OPERATING CHARACTERISTICS

*What receiver operating characteristics (ROCs) have been realized in either laboratory or field tests? What are typical implementation (e.g., scan and investigation) times?*

NCPA has collected A/S coupled mine detection data on more than 300 separate buried antitank and antipersonnel landmines and hundreds of square meters of clutter spots at U.S. Army field sites in Northern Virginia and Arizona. These sites have allowed for the technology to be tested under a wide range of environmental conditions. Blind tests of this technology, sponsored by the Army, have resulted in unprecedented results for the high probability of detection and low false alarms [4]. NCPA has provided A/S coupled mine detection data to the Army which has in turn been provided to university faculty in Florida and Missouri and to other Army contractors for the purpose of automatic target algorithm development. The use of these algorithms has resulted in almost perfect detection and zero false alarm rates [5]. The current implementation using a single LDV for data collection requires approximately two minutes to scan a square meter since it can only interrogate a single point on the ground at any given time.

## KNOWN OR SUSPECTED LIMITATIONS OR RESTRICTIONS ON APPLICABILITY

*Under what conditions (background clutter, mine type, environmental conditions, etc.) should the technology perform exceptionally well or poorly? What are the principal factors limiting current performance?*

The ideal operational condition for A/S coupled mine detection is a desert, sandy soil environment. The most serious limitation of the physics of A/S coupled buried mine detection is attenuation of the acoustic signal with depth, realistically limiting the detection depth to less than 30 cm. The technique is immune to moisture, weather, acoustic, and seismic noise sources as well as most natural clutter, including rocks and roots. Some man-made compliant objects, such as empty soda and paint cans, will emulate landmines. Based on winter testing in Alberta, Canada, hard frozen ground may also limit the capability of this sensor. From a measurement standpoint, detection speed is currently a limiting factor, as is the presence of heavy vegetation on the ground. Detection speed is limited by the measurement technology, which currently uses only a single LDV. Also, heavily covered grassy surfaces, particularly those with dead vegetation not directly rooted into the soil, present a challenge to the

LDV currently used to measure the A/S coupled vibrations at the ground's surface.

## ESTIMATED POTENTIAL FOR IMPROVEMENT IN THE TECHNOLOGY OVER TWO TO SEVEN YEARS

*Can the current limiting factors be overcome, and if so, what is the best realistic performance (in terms of ROCs and operational times) that could be expected?*

Considering that the ideas presented thus far have been funded and investigated for less than five years, the results are phenomenal. The most significant progress to be made will be in the area of the sensor used to measure the surface vibrations. Europe leads the world in the development of optical techniques for vibration sensing and the sensors used to date in the program are purchased from European countries. Significant increases in detection speed up to a 3-m swath at a few kilometers per hour may be accomplished through the use of multiple LDVs operating in parallel or development of alternative sensors. Potential alternative sensors for vibration measurement include Doppler focal plane array cameras, optical tilt cameras, Doppler radar, Doppler acoustics, and holographic speckle sensors. These alternatives face challenging sensitivity issues, and some, including Doppler radar and acoustics, have been able to sense the vibrations from mines in vegetation under laboratory conditions.

## OUTLINE OF A SENSIBLE RESEARCH AND DEVELOPMENT PROGRAM THAT COULD REALIZE THIS POTENTIAL, WITH ROUGH PROJECTED COSTS TO THE EXTENT POSSIBLE

The crucial elements of an A/S coupled mine detection research and development effort must include development of a continuously moving LDV array, further research into nonlinear acoustic phenomena, and investigation of alternative sensors. To further increase the probability of detection and reduce false alarms, NVESD is planning to fund an effort to fuse NCPA's A/S technology with a ground-penetrating synthetic aperture radar (GPSAR) developed by Planning Systems Inc. These two technologies are orthogonal in that they exploit disparate physical phenomena yet produce data in similar formats allowing fusion at the pixel level. They optimally operate under different conditions. Whereas A/S mine detection is limited to shallower depths, GPSAR works best at greater depths. A/S mine detection works best against plastic mines, while GPSAR works better against metallic mines [6]. Consequently, an ideal mine detection program would not only address the A/S issues but would also incorporate sensor fusion to include the GPSAR capability as well.

As mentioned above, the current mine detection effort is primarily focused on antitank mine detection at the direction of the sponsor. Because of the smaller size and multiplicity of shapes encountered in antipersonnel mine detection [7], a separate, independent program is advisable. NCPA estimates a need for approximately $3.5 million per year for four to five years to make this system ready for use.

## REFERENCES

1. N. Xiang and J. Sabatier, "Land Mine Detection Measurements Using Acoustic-to-Seismic Coupling," *Society of Photo-Optical Instrumentation Engineers,* No. 4038, 2000, pp. 645–655.

2. J. Sabatier and N. Xiang, "An Investigation of Acoustic-to-Seismic Coupling to Detect Buried Antitank Landmines," *IEEE Transactions on Geoscience and Remote Sensing,* Vol. 39, No. 6, June 2001, pp. 1146–1154.

3. M. Korman and J. Sabatier, "Nonlinear Acoustic Techniques for Landmine Detection: Experiments and Theory," *Proceedings of the Society of Photo-Optical Instrumentation Engineers AeroSense 2002 Conference,* 2002.

4. E. Rosen, K. Sherbondy, and J. Sabatier, "Performance Assessment of a Blind Test Using the University of Mississippi's Acoustic/Seismic Laser Doppler Vibrometer (LDV) Mine Detection Apparatus at A. P. Hill," *Society of Photo-Optical Instrumentation Engineers,* No. 4038, 2000, pp. 656–666.

5. P. Gader and A. Hocaoglu, "Continuous Processing of Acoustic Data for Landmine Detection," *Proceedings of the Society of Photo-Optical Instrumentation Engineers AeroSense 2002 Conference,* 2002.

6. M. Bradley, J. Sabatier, and T. Witten, "Fusion of Acoustic/ Seismic and Ground Penetrating Radar Sensors for Antitank

Mine Detection," *Society of Photo-Optical Instrumentation Engineers,* No. 4394, 2001, pp. 979–990.

7. N. Xiang and J. Sabatier, "An Experimental Study on Anti-Personnel Landmine Detection Using Acoustic-to-Seismic Coupling," *Journal of the Acoustical Society America,* No. 112, December 2002.

# ACOUSTIC/SEISMIC METHODS (PAPER II)

*Dimitri M. Donskoy, Stevens Institute of Technology*

## PHYSICAL PRINCIPLES

The essence of the acoustic/seismic approach is to excite low-frequency (typically below 1,000 Hz) vibration of a buried mine and measure surface "vibration signature" above the mine using remote sensors. Excitation of a mine and surrounding soil is achieved by using airborne (acoustic) and/or solid-borne (seismic) waves. Remote sensing is achieved with laser Doppler, microwave, or ultrasonic vibrometers.

The technique does not depend on the material from which the mine is fabricated, whether it be metal, plastic, wood, or any other material. It depends on the fact that a mine is a "container" whose purpose is to contain explosive materials and associated detonation apparatus. The mine container is in contact with the soil in which it is buried. The container is an acoustically compliant article whose compliance is notably different from the compliance of the surrounding soil. Dynamic interaction of the compliant container and the soil on top of it leads to specific linear and nonlinear effects used for mine detection and discrimination. The mass of the soil on top of a compliant container creates a classical mass-spring system with a well-defined resonance response. In addition, the connection between mass (soil) and spring (mine) is not elastic (linear) but rather nonlinear because of the separation of the soil/mine interface in the tensile phase of applied dynamic stress. These two effects, constituting the mine's "vibration signature," have been measured in numerous laboratory and field tests, which proved that the resonance and nonlinear responses of a mine/soil system can be used for detection and discrimination of buried mines. Thus, the fact that the mine is buried is turned into a detection advantage. Because the seismo-acoustic technique intrinsically detects buried "containers," it can discriminate mines from noncompliant false targets, such as rocks, tree roots, chunks of metal, bricks, etc. This was also confirmed experimentally in laboratory and field tests.

## STATE OF DEVELOPMENT

The technology is at the applied research stage. This consists of a considerable amount of laboratory research. The University of Mississippi and Stevens Institute of Technology have gone into the field to take data under semi-realistic conditions at Army test lanes. The Georgia Institute of Technology may soon initiate field tests as well. The University of Missouri, University of Florida, Ohio State University, SAIC, and Scientific Systems Company Inc. make efforts in the area of data processing and automatic target detection using seismo-acoustic data. MetroLaser Inc. has a program to build improved laser Doppler vibrometers, which would improve speed and sensitivity. Stevens Institute of Technology, in collaboration with Land Mine Detection System Inc., is in the process of developing an inexpensive microwave vibrometer/seismometer. The Army's Night Vision and Electronic Sensors Directorate has an in-house research program, funds most of the preceding organizations, and provides test facilities, coordination, and oversight.

Field testing is an integral part of the overall program and cannot be priced separately.

## CURRENT CAPABILITIES AND OPERATING CHARACTERISTICS

Excellent receiver operating characteristic (ROC) curves have been obtained against antitank mines. ROC curves do not exist against antipersonnel mines, although for some implementations very promising results were demonstrated.

Scan times are at present relatively slow: An off-the-shelf scanning laser Doppler vibrometer scans at discrete points with 50–100 m per dwell point. Antitank mines may require approximately 5–10 cm spatial resolution, while antipersonnel mines may require 1–2 cm resolution. Scan time is 6–45 seconds per square meter for antitank mines and 125–1,000 seconds per square meter for antipersonnel mines. However, several efforts are under way to greatly improve the speed. Specifically, an array of inexpensive sensors could increase scanning speed to at least an order of magnitude.

## KNOWN OR SUSPECTED LIMITATIONS

The technology is most sensitive to dynamically compliant mines. As a rule, nonmetallic mines are more compliant and easier to detect with seismo-acoustic detection.

To the extent that it has been tested, the technology is insensitive to most clutter and environmental conditions. While the technology has been shown to work in short grass, the use of a laser Doppler vibrometer prevents operation in moderate to heavy vegetation, and new types of sensors are needed to overcome this limitation.

The principal factor limiting current performance is limitation of existing sensing technology. Commercially available laser Doppler vibrometers cannot perform continuous scanning, do not provide adequate sensitivity because of speckle noise, exhibit unstable behavior in outdoor use because of environmental factors (temperature, humidity, etc.), and have inadequate laser power for soils with low reflectivity, degrading performance for oblique angles.

## POTENTIAL FOR IMPROVEMENTS

Potential improvements include increased operating speed (order of magnitude), improved sensor sensitivity and stability of operation under variable outdoor conditions, and development of vegetation-penetrating sensors to measure ground vibrations. Realistic ROC curves are expected to be excellent.

## OUTLINE OF A SENSIBLE RESEARCH AND DEVELOPMENT PROGRAM

A research program should address the following major tasks:

- sensor development to overcome limitations outlined above
- efficient acoustic/seismic energy delivery systems
- algorithms for data processing
- field testing and large-scale data collection.

The estimated cost of this program is $10–12 million per year for three to four years.

## BIBLIOGRAPHY

Costley, R. D., J. M. Sabatier, and N. Xiang, "Forward-Looking Mine Detection System," in *Detection and Remediation Technologies for Mines and Minelike Targets VI,* A. C. Dubey, J. F. Harvey, J. T. Broach, and V. George, eds., Seattle: International Society for Optical Engineering, 2001, pp. 617–626.

Donskoy, D. M., "Detection and Discrimination of Nonmetallic Mines," in *Detection and Remediation Technologies for Mines and Minelike Targets IV,* A. C. Dubey, J. F. Harvey, J. Broach, and R. E. Dugan, eds., Seattle: International Society for Optical Engineering, 1999, pp. 239–246.

Donskoy, D. M., "Nonlinear Vibro-Acoustic Technique for Land Mine Detection," in *Detection and Remediation Technologies for Mines and Minelike Targets III,* A. C. Dubey, J. F. Harvey, and J. Broach, eds., Seattle: International Society for Optical Engineering, 1998, pp. 211–217.

Donskoy, D. M., N. Sedunov, A. Ekimov, and M. Tsionskiy, "Nonlinear Seismo-Acoustic Land Mine Detection and Discrimination," *Journal of the Acoustical Society of America,* Vol. 111, No. 6, 2002.

Donskoy, D. M., N. Sedunov, A. Ekimov, and M. Tsionskiy, "Optimization of Seismo-Acoustic Land Mine Detection Using Dynamic Impedances of Mines and Soil," in *Detection and Remediation Technologies for Mines and Minelike Targets VI,* A. C. Dubey, J. F. Harvey, J. T. Broach, and V. George, eds., Seattle: International Society for Optical Engineering, 2001, pp. 575–582.

Sabatier, J. M., and N. Xiang, "Laser-Doppler Based Acoustic-to-Seismic Detection of Buried Mines," in *Detection and Remediation Technologies for Mines and Minelike Targets IV,* A. C. Dubey, J. F. Harvey, J. Broach, and R. E. Dugan, eds., Seattle: International Society for Optical Engineering, 1999, pp. 215–222.

Scott, W. R., Jr., C. Schroeder, and J. S. Martin, "Acousto-Electromagnetic Sensor for Locating Land Mines," in *Detection and Remediation Technologies for Mines and Minelike Targets IV,* A. C. Dubey, J. F. Harvey, J. Broach, and R. E. Dugan, eds., Seattle: International Society for Optical Engineering, 1999, pp. 176–186.

Scott, W. R., S. Lee, G. D. Larson, J. S. Martin, and G. S. McCall, "Use of High-Frequency Seismic Waves for the Detection of Buried Land Mines," in *Detection and Remediation Technologies for Mines and Minelike Targets VI,* A. C. Dubey, J. F. Harvey, J. T. Broach, and V. George, eds., Seattle: International Society for Optical Engineering, 2001, pp. 543–553.

Xiang, N., and J. M. Sabatier, "Anti-Personnel Mine Detection Using Acoustic-to-Seismic Coupling," in *Detection and Remediation Technologies for Mines and Minelike Targets VI,* A. C. Dubey, J. F. Harvey, J. T. Broach, and V. George, eds., Seattle: International Society for Optical Engineering, 2001, pp. 535–541.

Xiang, N., and J. M. Sabatier, "Land Mine Detection Measurements Using Acoustic-to-Seismic Coupling," in *Detection and Remediation Technologies for Mines and Minelike Targets V,* A. C. Dubey, J. F. Harvey, J. Broach, and R. E. Dugan, eds., Seattle: International Society for Optical Engineering, 2000, pp. 645–655.

Appendix I

# ELECTRICAL IMPEDANCE TOMOGRAPHY

*Philip Church, Neptec Design Group*

## BASIC PRINCIPLES

Electrical impedance tomography (EIT) is a technology developed to image the electrical conductivity distribution of a conductive medium. The technology is of interest because of its low cost and also because the measurement of the electrical conductivity brings direct information about the composition of the conductive medium. Because the ground is conductive to a certain extent, the technology can also be used to detect buried objects. The application of landmine detection is of particular interest because the object is usually buried at shallow depths and causes a discontinuity in the soil conductivity that can be sensed from the surface of the ground.

EIT uses low-level electrical currents to probe a conductive medium and produce an image of its electrical conductivity distribution. While a pair of electrodes is stimulated, the electrical voltage is measured on the remaining pairs of electrodes. After all the independent combinations of interest have been stimulated, an algorithm using the measured data reconstructs an image of the electrical conductivity distribution within the volume. In the case of ground probing, an array of electrodes is placed on the surface of the ground to provide an image of the conductivity distribution below the surface. The EIT technology will detect mines buried in the ground by detecting ground conductivity anomalies. The presence of a metallic or non-conductive mine will disturb the conductivity distribution in the soil. The signal characteristics are based on the size, shape, conductivity, and depth of the buried mine.

Figure I.1 shows an EIT detector prototype optimized for antitank landmines. A typical EIT detector has three major components: the electrode array, the data acquisition system, and a data processing unit. In this case, the electrode array comprises 8 columns and 8 rows of electrodes—for a total of 64 electrodes. The electrodes are spring-loaded and can adjust with the terrain variations. The data acquisition system incorporates the electronics and firmware required for the electrical stimulation of the electrodes and the recording of the resulting potentials. Typically the stimulation current is on the order of 1 mA and the frequency of the stimulation is about 1 kHz. The data processing unit is a software application that processes the raw measurements using a mine detection algorithm based on a matched filter approach. The detector response is precalculated for a replica of the size and shape of the object of interest—for a number of grid locations underneath the detector. A correlation is then performed between the detector response for the replica and the actual detector response obtained from the measurements, for all the replica positions considered. The position that yields the largest correlation value is identified as the most likely position for the mine. Figure I.2 shows an example of the detector

RANDMR1608-I.1

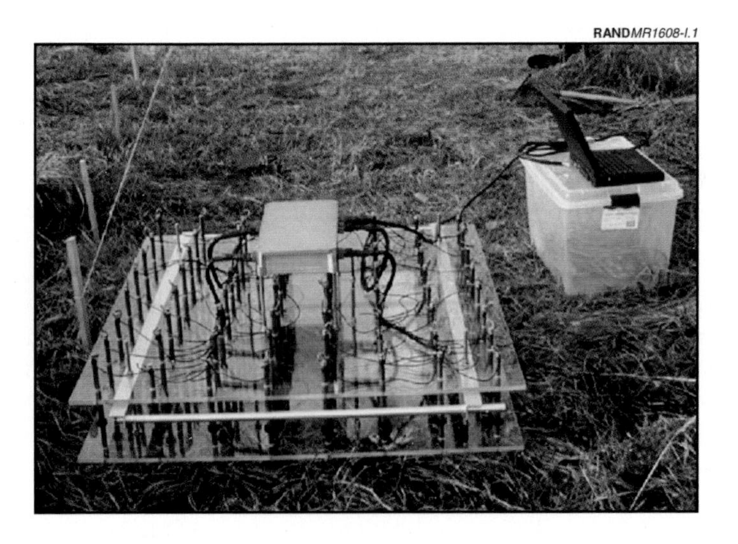

**Figure I.1—EIT Landmine Detector Prototype**

response for a nonconductive mine-like object buried at a depth of 14 cm in a sandy soil. The three-dimensional graph represents the detector response as a function of the positions in the x-y plane, in units of meters. The detector provides a similar response for a metallic object, with a sign reversal.

The prototype shown in Figure I.1 was built in view of evaluating the EIT technology as a confirmatory detector for antitank mines. A smaller lab unit was also built, suitable for objects with a size typical of antipersonnel landmines.

## STATE OF DEVELOPMENT

The EIT technology is relatively recent and has been researched mostly for medical diagnostic applications. The research done in the application of EIT to detect landmines has been very limited. Under the sponsorship of Defence R&D Canada–Suffield (DRDC-Suffield), research has been conducted to assess the EIT technology capability

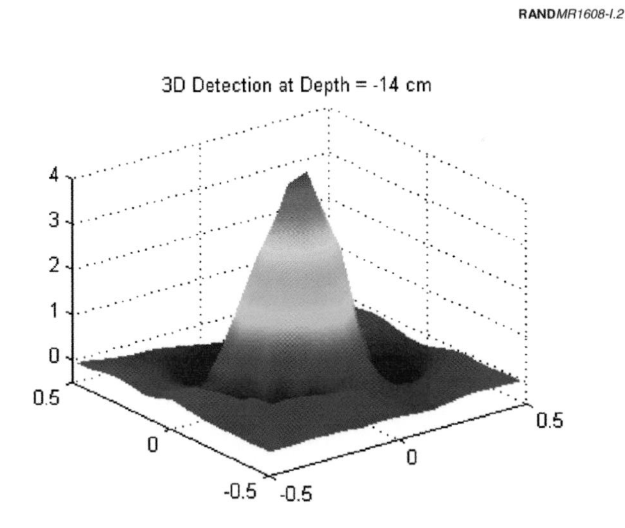


**Figure I.2—Detector Response for an Antitank Mine-Like Object Buried at a Depth of 14 cm**

to detect, first, buried unexploded ordnance (UXO) and, later, buried antitank mines. Wexler was one of the first investigators to apply EIT to the detection of UXO in the ground [1,2]. The detection of UXO requires probing depths that are large relative to the size of the surface electrode array and EIT may not be the best technology for this particular application. The EIT technology is sensitive to noise, which prevents its use at depths that are large with respect to the electrode array size.

It holds promise, however, when the problem is constrained to the detection of objects buried at shallow depths. The author of this report, in conjunction with John McFee of DRDC-Suffield, has recently developed a prototype EIT detector aimed at antitank landmine detection. Additional details on the detector design and performance are reported in previous reports by the authors [3,4].

An amount of approximately $400,000 has been invested so far in the evaluation of that technology for landmine detection. A prototype EIT detector has been built and evaluated in a laboratory environment and in a limited set of field trials. Further evaluations are required to assess its performance in various environmental conditions.

## CURRENT CAPABILITIES

### Detection Performance

Currently, insufficient data exist to derive a statistically meaningful set of receiver operating characteristic (ROC) curves. The detector has been evaluated in soils of conductivity varying from 1 mS per meter to 15 mS per meter. The detector performed well in detecting antitank mines and antitank mine-like objects down to a depth of 15–20 cm, where the depth is measured from the top of the object to the surface of the ground. As a general rule, reliable detections were obtained down to a range of 1.0–1.5 electrode spacings for objects with a size on the order of two electrode spacings. The matched filter approach is also very efficient at reducing the false alarms caused by objects of different sizes.

### Data Acquisition and Processing Time

The acquisition and processing time is determined by the number of independent configurations of stimulating and recording pairs of electrodes. For a 64-electrode detector, there are approximately 2,000 independent configurations. The data acquisition time, using currently available data acquisition electronics, is on the order of 1 second for a complete scan. The core of the data processing developed for the current prototype is based on a matched filter implemented through matrix operations. This takes a few seconds in the Matlab™ tool environment but would take only a few milliseconds on a dedicated processor.

## STRENGTHS AND LIMITATIONS

### Strengths of the Technology

In a mine detection application, EIT technology has the following strengths:

**Metallic and nonmetallic landmines.** Because the EIT technology detects a perturbation in the local soil electrical conductivity, it does not matter whether the perturbation is caused by a conductive or nonconductive object.

**Performance in wet areas.** The EIT technology appears to have a special niche in wet environments, such as beaches, ocean littorals, rice paddy fields, marshes, and other wet areas, because this is where EIT works at its best, mostly because the environmental wetness ensures a good electrical contact. The detector has proven to be unexpectedly efficient in sand, even if the sand is poorly conductive, as long as it holds a bit of moisture. Currently, very few detectors can detect nonmetallic mines buried in wet, conductive areas.

**Low cost.** The hardware required to build an EIT detector is simple and of relatively low cost.

### Limitations of the Technology

In a mine detection application, EIT technology has the following limitations:

**Electrode-soil contact.** The EIT detector requires an electrical contact between the electrode array and the soil. Electrical contact cannot be ensured in some environments, and the deployment of electrodes in close proximity to explosives is a potential operational issue, although no large force is required to achieve the electrical contact. The use of the detector in a water environment eliminates the need for a direct contact because the electrical conduction is achieved through the water medium.

**Environment.** The EIT detector requires an environment that is electrically conductive in order that it work properly. The detector cannot work properly in such environments containing dry sand or rock covered surfaces.

## POTENTIAL FOR IMPROVEMENT

There is significant room for the improvement of the EIT technology applied to landmine detection. In particular, further work would be especially useful in the areas of environmental evaluations, deployment, and instrumentation. This is discussed in more detail below.

**Environmental evaluations:** Elaborate field trials are required to better understand the performance of an EIT detector in terms of ROC curves, in a variety of environments. The EIT technology is showing a significant potential for the detection of mines in wet environments. Further evaluations should be carried out in environments, such as beaches and marshes, where this detection approach can make a difference with respect to other landmine detection modalities.

**Deployment platforms:** Further work is required to design and evaluate deployment platforms for environments where the EIT technology is especially promising. Conceptual deployment platforms include, for example, remote-controlled rovers equipped with electrodes embedded in tracks to go over beaches or flexible mats with embedded electrodes to be dragged on the bottom of a body of water.

**Instrumentation improvements:** Current technology allows for data acquisition and processing on the order of 1 second. This could be improved with faster electronics and dedicated processors. The development and evaluation of fast algorithms capable of identifying a conductivity perturbation in environments having a complex conductivity distribution, such as multiple conductivity layers, is also of interest.

## SUGGESTED RESEARCH AND DEVELOPMENT PROGRAM

It is suggested that a research and development (R&D) program aimed at advancing the EIT technology for mine detection applications should include the items identified above as "potential for improvement." A rough order of magnitude (ROM) cost estimate to perform such an R&D program is provided in Table I.1.

**Table I.1**

**Suggested R&D Program**

| R&D Component | ROM Cost | Description |
|---|---|---|
| Environmental evaluations | $1–2 million | Field trial evaluations in environments for which the EIT technology is especially well suited, such as beaches and marshes.<br>• Preparation and execution of the field trials<br>• Analysis of the data<br>• Preparation of a report and recommendations |
| Deployment platforms | $5–10 million | Design and build prototype platform(s) that are adapted for the environments where the EIT performs best. |
| Instrumentation improvements | $2–4 million | Improvements to data acquisition electronics and data processing algorithms. |

## REFERENCES

1. A. Wexler, "Electrical Impedance Imaging in Two and Three Dimensions," *Clinical Physics and Physiological Measurement*, Supplement A, 1988, pp. 29–33.

2. A. Wexler, B. Fry, and M. R. Neuman, "Impedance-Computed Tomography Algorithm and System," *Applied Optics*, No. 24, December 1985, pp. 3985–3992.

3. P. Wort, P. Church, and S. Gagnon, "Preliminary Assessment of Electrical Impedance Tomography Technology to Detect Mine-Like Objects," in *Detection and Remediation of Mines and Mine-like Targets IV*, A. C. Dubey, J. F. Harvey, J. Broach, and R. E. Dugan, eds., Seattle: International Society for Optical Engineering, 1999, pp. 895–905.

4. P. Church, P. Wort, S. Gagnon, and J. McFee, "Performance Assessment of an Electrical Impedance Tomography Detector for Mine-Like Objects," in *Detection and Remediation Technologies for Mines and Minelike Targets VI*, A. C. Dubey, J. F. Harvey, J. T. Broach, and V. George, eds., Seattle: International Society for Optical Engineering, 2001, pp. 120–131.

Appendix J

# NUCLEAR QUADRUPOLE RESONANCE (PAPER I)

*Andrew D. Hibbs, Quantum Magnetics*

## INTRODUCTION

Nuclear quadrupole resonance (NQR) combines the spatial localization capability and convenience of metal detection or ground-penetrating radar (GPR) with the compound specific detection capability offered by chemical detection techniques. Starting in the mid-1990s, groups at Quantum Magnetics Inc. and the Naval Research Laboratory made considerable improvements in the basic scientific and instrumentation techniques for performing an NQR measurement over the ground. These advances have led to programs to build a vehicle-mounted NQR system to work in conjunction with the Army Ground Standoff Mine Detection System (GSTAMIDS) mine clearance system, and a man-portable system to meet Marine Corps requirements for a man-carried mine detector. Overall, NQR is the only new technology in the past 10 years to progress to the stage at which practical deployment has become a possibility.

NQR is an electromagnetic technique–based operating system in the frequency range of 0.5–5.0 MHz. Its current limitations for landmine detection are

- Insufficient signal for rapid detection of small deeply buried TNT mines
- Radio frequency interference; in particular, commercial AM radio stations within 20 kHz of the frequencies for detection of TNT, and electromagnetic noise sources in the immediate vicinity of the system.

To develop the full potential for NQR landmine detection, a number of ancillary issues must be addressed:

- Development of techniques to enable an NQR system to operate while in motion. In its present form, an NQR detection coil can operate only while stationary.
- Development of array techniques in order to operate multiple NQR detection coils in close proximity of each other.
- Reduction of the overall system size, weight, and power to enable integration of an NQR with other mine detection sensors on a single platform.
- Engineering modifications to improve reliability and reduce system cost.

There are two general approaches to deploying NQR technology for landmine detection. The first is to work to solve the basic limitations (e.g., the first items of the lists above) in order to build a stand-alone NQR system operating in a conventional sweeping mode detecting all types of mines. The principal risk is that some problems, item 1 in particular, may be difficult to solve completely. The second approach is to focus on using NQR where it is presently strong, such as in the detection of RDX and CompB mines, or operating in a stationary mode as a confirmation sensor to augment other mine detection technologies. This appendix will address the results that can be expected in taking each of these approaches.

## MAGNITUDE OF THE NQR SIGNAL FOR THE EXPLOSIVES USED IN LANDMINES

In contrast to GPR, metal detection, and many other technologies that are primarily limited by ground clutter, NQR technology is usually limited by its own internal system noise. This noise is thermal, which means that extending the interrogation time, t, increases the overall signal-to-noise ratio (SNR) of the measurement by $\sqrt{t}$ (for example, doubling the measurement time increases the SNR by a factor of 41 percent). Cancellation of radio frequency interference (RFI) appears to scale as $\sqrt{t}$ also, and so its effect can be included by a fixed decrease in the system SNR. Given that the NQR signal is proportional to the mass of explosive, the most suitable way to describe the performance of an NQR system is the SNR per unit time per unit mass of explosive. The present SNR for the present man-portable NQR system on 100 g of the three types of explosive used in antipersonnel mines deployed at tactical depths with zero RFI for a 1-second measurement is listed in Table J.1.[1]

Because the statistics of the NQR system noise are well defined, the relationship between the probability of detection (PD) and probability of false alarm (PFA) for a given measurement SNR can be accurately calculated. PD/PFA pairs at operating points relevant to mine detection are shown in Table J.2.

For example, these results show that an NQR measurement on a small RDX mine in an RFI environment that results in a 20-percent loss in SNR (the present target for the antitank detection systems) provides an SNR of 4.7, which is easily large enough to meet present performance goals. The worst case at the present time is the TNT antipersonnel mine, for which a 1-second measurement is inadequate even under conditions of zero RFI. To combat this, four individual measurements are necessary to detect such mines. Unfortunately, a property of NQR for TNT (not shared by RDX and only in a

**Table J.1**

**Present Signal-to-Noise Ratio for a 1-Second NQR Measurement of 100 g of Explosive**

| | 100 g RDX | 100 g TNT | 100 g Tetryl |
|---|---|---|---|
| SNR in 1 second | 13.1 | 1.2 | 10.7 |

**Table J.2**

**Example PD and PFA Operating Points 2 to 7 for NQR SNR**

| | Operating Point/SNR | | | | | |
|---|---|---|---|---|---|---|
| | 2 | 3 | 4 | 5 | 6 | 7 |
| Probability of detection (%) | 88.00 | 91.00 | 95.00 | 99.60 | 99.80 | 99.99 |
| Probability of false alarm (%) | 20.0 | 5.0 | 1.0 | 1.0 | 0.1 | 0.1 |

[1]The NQR signal is also temperature dependent and varies a little with ground conductivity. The data are at 20°C and are for typical ground conditions.

minor form by tetryl) is that immediate repeat measurements are not possible. Instead, a dead time on the order of 5–10 seconds is required between TNT measurements (e.g., the optimum time for four measurements at 20°C is 20 seconds). The need for this dead time considerably complicates the use of NQR to detect antipersonnel mines that contain only TNT.

## PERFORMANCE PROJECTIONS FOR NQR AS A STAND-ALONE MINE DETECTOR

To date, two separate mine detection configurations have been pursued: large vehicle-mounted coils for road clearance applications and minimal weight systems for man-portable use. The size and weight of NQR technology is comparable to GPR and metal detection systems, and so these two categories naturally apply to NQR systems as well. It should be noted that NQR per se cannot be used to detect metal-cased mines because of shielding of the applied RF field. However, a metal-cased mine is easily detected by a low-grade metal detector and tests have shown that such mines can also be detected by their electrical loading effect on the QR coil. Furthermore, it is believed that a stand-alone NQR system could be easily modified to detect metal-cased mines at a level greater than 99-percent PD, and so such mines are excluded from this study. To include such mines in the analysis below, the projected PDs should be increased by a weighting factor based on the respective ratio of mine types.

### Performance of Current NQR Technology as a Stand-Alone Sensor

As indicated, NQR is adequate to detect RDX and tetryl-based antipersonnel mines at present tactical depths and for present mine sizes—but not for small TNT mines. However, not all mines are small and contain TNT, and it is of practical interest to consider a spectrum of explosive types, explosive masses, and mine burial depths. Computer simulations using outdoor experimental results as calibration have been written to calculate the average PD for the type of mines contained in the Army test lanes at Yuma (Ariz.) and Aberdeen (Md.) proving grounds, over the full range of expected temperature and mine depths.[2] The results are listed in Table J.3 for the man-portable NQR system presently under development. Also included, for reference, is the PD on antitank mines for the man-portable system and also for the vehicle-mounted NQR system currently under development. The present hardware for these two systems is shown for reference in Figure J.1. Each system detects mines roughly within the footprint defined by the diameter of the detection coil. The vehicle-mounted system is designed to detect antitank mines only. However, antitank mines are a potential threat in humanitarian demining, and, in addition, the larger NQR detection coil for the vehicle-mounted system could be split into a number of smaller coils to detect antipersonnel mines with a larger area per measurement, thereby increasing clearance rate.

The projections shown in Table J.3 are fully supported by previous tests[3] and show that NQR has adequate sensitivity in its present form to provide a threshold level of sensitivity. However, these results should be viewed with the following caveats, all of which affect the system SNR and highlight the need for continuing development efforts in NQR:

1. All NQR systems to date have been operated with a stationary detection coil. Thus the present systems must be moved from

**Table J.3**

**Projected Results for Detection of Low-Metal Landmines at Aberdeen and Yuma Proving Grounds (Net PD at 5-Percent PFA Over All Mine Types, Burial Depth and Temperatures)**

| | Low-Metal AP Mines (%) | Low-Metal AT Mines (%) |
|---|---|---|
| Man portable | 96.7 | 94.9 |
| Vehicle mounted | N/A | 99.5 |

[2]The variation of SNR with mine depth is very similar to that for metal detectors. Based on experimental data, ground conductivity is taken to be 50 mS per meter for the man-portable system and 20 mS per meter for the vehicle mounted.

[3]The vehicle-mounted NQR system was formally tested at Yuma Proving Ground in April 2002. The final test results have not yet been made available for release. In a dry run, in February 2002, the system recorded 100-percent PD (63/63).

RANDMR1608-J.1

**Figure J.1—Present Configuration of the Man-Portable (left) and Vehicle-Mounted NQR Systems (right)**

point to point, stopping to carry out the measurement. Preliminary experiments show that the NQR signal for a moving system is about a factor of 2 lower than that for a stationary system and that this signal decreases by about 10 percent going from velocities of 1 to 3 m per second.

2. The projections for the present performance are for a coil-to-ground standoff of 2.0 cm and antipersonnel mines at depths of up to 7.5 cm. This is comparable to the present use of the Handheld Standoff Mine Detection System (HSTAMIDS) currently being developed by the Army, but the NQR signal decreases more with increased standoff than GPR. Increasing the separation from the mine to the NQR by 1 cm reduces the SNR by a factor of approximately 30 percent.
3. The analysis assumes that the background RFI level is mitigated to within 1 decibel of the coil noise floor. There is an insufficient amount of field data at this time to know to what extent the 1-decibel target can be met, but the RFI level is generally low in regions where humanitarian demining is needed. Note that the question of RFI for NQR is analogous to the problem of clutter due to metal fragments and rocks for metal detection and GPR, respectively. It is difficult to develop a test protocol that adequately assesses general performance against clutter.

The total system interrogation time is dominated by the need to detect TNT and is essentially 20 seconds for the cases in Table J.3 (RDX and tetryl can be probed during the dead time between TNT measurements). The coil for the man-portable NQR system covers approximately a 100-sq-cm region, and therefore the area scan rate for that system is approximately 30 minutes per square meter. However, *a critical advantage of NQR, particularly for humanitarian demining, is that because the SNR increases with measurement time, an NQR system can be used to reexamine its own alarms.* For example, 5-percent PFA translates to 5 false alarms per square meter for the man-portable system. These alarms could be reinterrogated in an additional 100 seconds, resulting in a notional system false alarm rate (FAR) of 0.25 per square meter with a small reduction in overall system PD (e.g., down to 93.5 percent for low-metal antipersonnel mines). This property means that best- and worst-case field conditions can generally be accommodated by varying the NQR measurement time.

One possible scenario with present technology would be to increase the basic NQR measurement time by a factor of 2, which would increase the PD for the man-portable case shown in Table J.3 to 99.9 percent and 99.6 percent for antipersonnel and antitank mines, respectively, in a time on the order of 1 hour per square meter. The 5 false alarms per square meter would be remeasured in an additional 200 seconds for a total scan time of 63 minutes per square meter with a PD of greater than 99 percent. Alternatively, to reduce scan time, the NQR system could be operated as a confirmation sensor for essentially any other mine detection technology. One example of operation in this mode is discussed in the Combination of NQR with Other Mine Detection Technologies section below.

## Performance Projections for Future Stand-Alone NQR Landmine Detection Systems

To develop a general-purpose NQR landmine detector to replace present metal detectors, two elements are needed: an increase in SNR of about a factor of 2 for stationary and 4 for in-motion detection of TNT and the development of hardware to use NQR while in motion. Given the need for improved TNT sensitivity, there has been essentially no work on the motion problem.

The majority of progress in improving the TNT signal in the past five years has come from increasing the power used and careful optimization of the measurement parameters. Further improvements in these specific areas are unlikely. Recently, a method has been developed to reduce the dead time needed between TNT measurements by a factor of 2. For the 20-second composite TNT measurement described above, this improvement increases the SNR by $\sqrt{2}$, but it does not increase the SNR per individual measurement and thus has no impact on the issue of using NQR in motion. A factor of $\sqrt{2}$ increase in SNR for the present man-portable configuration would increase the PD for antipersonnel and antitank mines to 99.9 percent and 99.6 percent, respectively (i.e., the same improvement as for doubling the measurement time). It is straightforward to predict the system performance for any given improvement in TNT sensitivity.

## COMBINATION OF NQR WITH OTHER MINE DETECTION TECHNOLOGIES

At present, NQR is being developed for use as a confirmation sensor to be used in conjunction with the Army HSTAMIDS and GSTAMIDS systems to reduce the overall system FAR. As such, it is already being designed for use in combination with other technologies, and a confirmation sensor role is ideal to NQR at its current state of development. However, a more intriguing possibility is to develop a truly integrated measurement approach in which the capabilities of NQR are used to relax the requirements on the initial radar and metal detection sensors and increase the PD of the initial primary scan.

### Approaches That Use Current NQR Technology

One possibility for a high-speed mine detector with high PD is to modify the detection algorithm for the HSTAMIDS system to increase PD and use NQR to cope with the associated increase in FAR. Specifically, GPR systems tend to have difficulty distinguishing shallow mines from ground surface effects but distinguish much better with deeper mines, while metal detectors detect shallow targets better but lose signal with depth. The present HSTAMIDS system takes the logical AND function of the GPR and metal readings to reduce FAR. However, if the logical OR is taken of the two HSTAMIDS sensors, then the PD increases to over 99 percent. Table J.4 compares

AND and OR HSTAMIDS modes when used in conjunction with an NQR confirmation sensor for the most difficult combination of low-metal mines in off-road conditions. If metallic mines and less demanding test conditions are included, the average PD for a combined NQR HSTAMIDS system is predicted to be in excess of 99.5 percent. Note that for this application it is not essential that the NQR system be physically integrated with the other sensors, but it could be used as a separate detection system.

In addition to sensor combinations that increase performance, one innovative approach would be to build a low-cost NQR system that could be placed over a suspect region and left to interrogate the ground for up to several minutes. Such a system could have extremely high PD and very low PFA (that could be set by the user) and could essentially replace the act of ground probing and excavation, thereby increasing area coverage rate and greatly improving safety.

## Projections for Future Multisensor Landmine Detection Systems That Include NQR

Future improvements in the NQR signal from TNT can be translated to improved detection performance along the lines illustrated above in Performance Projections for Future Stand-Alone NQR Landmine Detection Systems. However, a more significant advance would be to use NQR in an integral role as part of the primary mine detection function. One possibility is to use NQR to detect only RDX-based mines for a sweeping sensor and develop a combined system in

**Table J.4**

**Comparison of Two Possible Combinations of NQR with HSTAMIDS for Low-Metal Mines Measured in Off-Road Conditions**

| | PD (%) | FAR | Sweep Rate ($s/m^2$) |
|---|---|---|---|
| HSTAMIDS alone | 92.3 | 0.25 | 60 |
| HSTAMIDS (conventional mode) + NQR confirming | 97.0 | 0.09 | 74 |
| HSTAMIDS (logical OR mode) + NQR confirming | 99.3 | 0.28 | 105 |

which other sensors detect TNT mines. This has the obvious benefit of overcoming the present limited detection capability for TNT. One approach is to note that RDX-based mines are smaller and tend to have more complex shapes than TNT mines. As a result, present GPR systems have lower PD on such RDX antipersonnel mines than the less sophisticated TNT-based targets. Thus, one potential combination of sensors is to use an NQR system that detects only RDX in conjunction with a GPR system with an increased detection threshold. By raising its threshold the FAR of the GPR system could be reduced, possibly quite significantly. An NQR confirmation sensor could also be used to optimize the overall system PD, FAR, and scan time.

## SUMMARY

NQR provides a compound specific detection capability and combined with rescanning offers the opportunity for essentially zero false alarms. Present NQR technology has adequate PD/FAR to be used alone provided sufficient scan time is available. The required time could be reduced significantly if the TNT signal could be improved. In addition, a number of such practical issues as RFI have not yet been fully evaluated, and continued research and testing is needed in these areas.

An NQR confirmation sensor based on present technology can provide significant benefit when used to reinterrogate areas flagged by the existing mine detection systems. In this mode, NQR would replace the dangerous and time-consuming activity of ground probing and excavation. In addition, present NQR performance is adequate to consider integrating NQR more fully with present sensors, for example, by modifying the detection protocol used in combined radar and metal detection systems.

Finally, use of NQR as a primary sensor requires development of technology to enable an NQR system to operate while in motion. A system that detects only RDX is probably feasible with a moderate engineering effort, while a complete NQR primary sensor requires a significant increase in TNT sensitivity. While challenging, these problems have received little attention in the relatively brief time spent on NQR to date and may be solvable at moderate research and development costs.

Appendix K

# NUCLEAR QUADRUPOLE RESONANCE (PAPER II)

*Allen N. Garroway, Naval Research Laboratory*[1]

## INTRODUCTION: NUCLEAR QUADRUPOLE RESONANCE AT THE NAVAL RESEARCH LABORATORY

The Polymer Diagnostics Section (Code 6122) of the U.S. Naval Research Laboratory (NRL) has expertise in solid state nuclear magnetic resonance (NMR), magnetic resonance imaging (MRI), and nuclear quadrupole resonance (NQR). In 1983, NRL established an interagency agreement with the Federal Aviation Administration (FAA) to advise the FAA on NMR methods to detect explosives. We were aware of earlier work [1] on the use of NQR for explosives detection, and in 1987 we initiated a research program at NRL to explore NQR, with initial support of the FAA and Technical Support Working Group (Department of Defense). NQR explosives detection technologies, including methods applicable for landmine detection, have been developed and patented by NRL and have been licensed by the U.S. Navy to Quantum Magnetics (San Diego, Calif.), a subsidiary of InVision Technologies Inc.

## PHYSICAL PRINCIPLES

### Limitations of Conventional Detection Methods[2]

The basic technology for both military and humanitarian mine detection is still the electromagnetic metal detector, a direct descen-

[1]This work was prepared by a U.S. government employee as part of his official duties.

[2]This section is adapted from Garroway et al. [1].

dant of those used in World War II. Finding a metal-encased antitank mine (5–10 kg of explosive) buried 10 cm underground is trivial for such a device, but finding a "low-metal" antipersonnel mine (50–100 g of explosive, and perhaps 0.5 g of metal for the firing pin) below the surface is highly challenging. The electromagnetic return signal from the antipersonnel mine is much weaker, and so the operator must turn up the detector gain. At higher sensitivity, however, much more of the other metal detritus, such as nails and shell fragments, becomes visible to the detector. The operator is compelled to operate at very high sensitivity and to flag any alarm as a potential landmine. Unfortunately, the next step is the most difficult: One must then separate the landmines from the false alarms arising from this benign background of signals. Currently, that "resolution of false alarms" is still done by mechanical probing: The deminer or combat engineer performs very delicate archeology with a pointed stick to classify the source of the electromagnetic signal—a landmine, perhaps rigged with an antihandling device, or just a rusty nail. In that sense, finding landmines is easy; however, separating them from the clutter is tough—and extremely hazardous.

**Why NQR?** What is desired is a detector based on a signal that is specific to the landmine. Certainly a unique signature of the explosive would provide a way to reduce this clutter problem. Such arguments lead to chemical detection of the explosives in landmines. While dogs are being used to find landmines, their method of detection is still a subject of controversy and their efficiency is not high. The vapor pressure of the military explosives used in landmines is quite low, and commercially manufactured mines hermetically seal the explosives in a polymeric case. Further, as explosive vapors and particles are quite sticky, the transport of explosive from the main charge, through the case, and then through the ground is slow and inefficient. Vapor sensors have been explored for landmine detection, and there is a recent indication that, under some field conditions, exquisitely sensitive vapor detectors can detect the plume from a landmine [2].

But there is another method—NQR—that is specific to the chemistry of an explosive, regardless of how it is packaged.

## BASICS OF NQR

NQR [3] is a magnetic resonance phenomenon related to NMR and its offspring, MRI. In NMR and MRI, a large static magnetic field (0.05–20.00 T, 0.5–200.0 kG) orients the nuclei so that slightly more are in the low energy state (aligned parallel to the static field) than are in the higher state (opposed to the field). This population difference corresponds to a weak diamagnetism of the nuclear spins, with a classical magnetization vector aligned along the static magnetic field. The magnetic field corresponding to this nuclear diamagnetism can be observed by applying a resonant radio frequency (RF) pulse (at the Larmor frequency and at right angles to the static field), causing the magnetization to rotate away from the axis of the static magnetic field. The magnetization then precesses freely in the static field, at the Larmor frequency, and this time-dependent flux induces a weak voltage in an RF pickup coil perpendicular to the static field. This induced signal is the NMR signal.

### Comparison to Nuclear Magnetic Resonance

NQR is similar to NMR but has some important distinctions. In NQR, the splitting of the nuclear spin states is determined by the electrostatic interaction of the nuclear charge density, $\rho(r)$, with the external electric potential, $V(r)$, of the surrounding electron cloud (see Figure K.1). A moment expansion of this electrostatic interaction shows that the important coupling is between the nuclear quadrupole moment, indicated schematically in Figure K.1, and the second derivative of the electric potential (equivalently, the gradient of the electric field). *This is a key result.* The quadrupole moment, nonzero only for nuclei with spin quantum number I greater than or equal to 1, is a nuclear physics parameter describing the distribution of charge in the nucleus. (For landmine applications, the primary nucleus of interest is $^{14}N$, with I equal to 1.) The second term, the coupling to the electric field gradient of the valence electrons is largely based on chemistry, although the local crystal packing also plays a role.

Contrast the chemical specificity of NQR with that of NMR. While NMR provides highly detailed information about chemical structure, the range of "chemical shifts" is generally small. Hydrogens in any

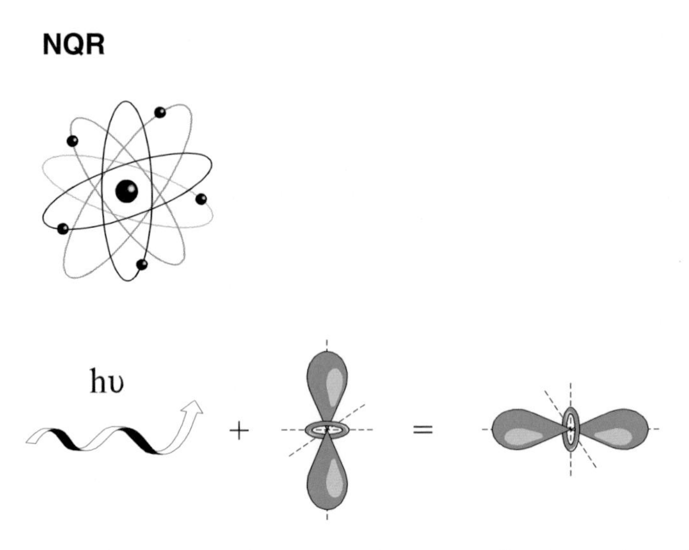


NOTE: Applying a pulse of the correct frequency υ flips the nuclear spin and induces an NQR signal in a pickup coil.

**Figure K.1—A Quadrupolar Nucleus Slightly Aligned by the Electrostatic Interaction with the Valence Electrons**

arbitrary organic structure differ by a range of about 10 ppm away from their nominal NMR frequency, e.g., a 6-kHz range of frequencies in a 600-MHz NMR spectrometer. However, for $^{14}N$ NQR, the NQR frequencies can range from *zero to 6 MHz,* depending on the symmetry of the molecule. Indeed, one of the difficulties of NQR is that it can be *too sensitive* to the chemistry of the compound of interest.

There are also some significant subtleties in NQR compared with NMR: For NQR the nuclear spin is greater than or equal to 1, and the spins are quantized along the principal axis system of the electric field gradient, rather than for the NMR or MRI case that (commonly) involves spin-1/2 nuclei quantized along the static magnetic field. For the present purpose, these distinctions are best overlooked, and, in a rough sense, it suffices to regard NQR as NMR without the magnet.

## NQR as a Detector

One significant advantage of NQR is the absence of a magnet: Even if the NMR approach were thought to give some advantage to detecting explosives, projecting a large static magnetic field into the ground is difficult. But the main advantage is that NQR provides a highly specific and arguably unique frequency signature for the material of interest. Figure K.2 shows the NQR frequencies for a number of common explosives, as well as for some narcotics and other materials. Although the chemical structure of RDX (Figure K.3) indicates that the three ring nitrogens are chemically equivalent and hence would be expected to have identical NQR frequencies, in fact the crystal packing is sufficient to remove this degeneracy, and indeed the chemically equivalent ring nitrogens are separated by the order

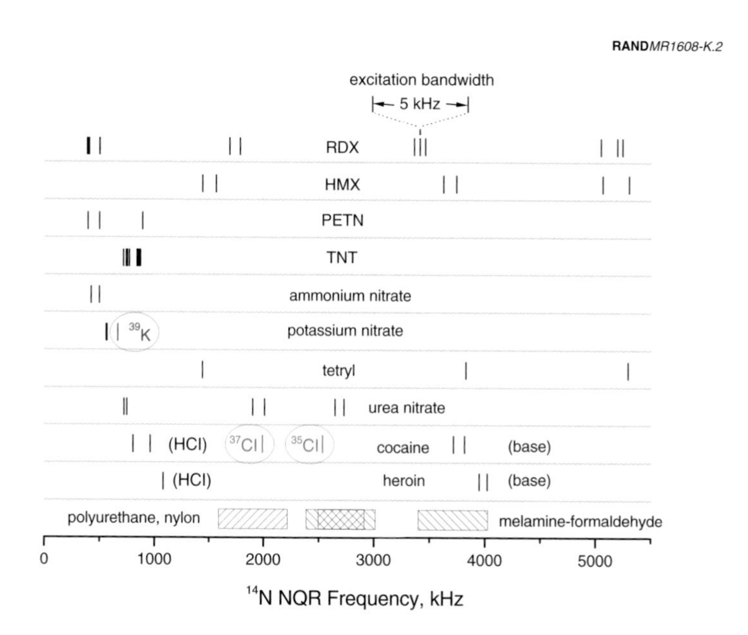


NOTE: Unless otherwise indicated, the frequencies are for $^{14}N$. Also shown are estimates for the general frequency bands expected for some commercial polymers.

**Figure K.2—Representative NQR Frequencies for Some Explosives and Narcotics**

RANDMR1608-K.3

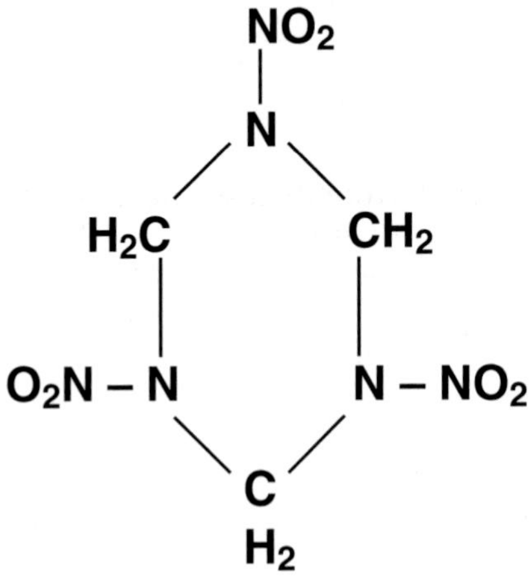

**Figure K.3—Chemical Structure of RDX**

of 100 kHz from one another (see Figure K.2). This demonstrates the specificity of NQR: Even such small effects from crystal packing are sufficient to resolve the NQR lines from nominally equivalent nitrogens. Because the bandwidth of excitation is only about 5 kHz for commercial NQR detectors, NQR lines more than 5 kHz away from the carrier will not be excited.

For landmine detection, TNT, RDX, and, to a lesser extent, tetryl are the most important explosives. The basic detection concept is particularly simple: Apply a pulse or series of RF pulses resonant at the appropriate NQR frequency of the explosives of interest, and look for the presence (or absence) of a return signal.

As discussed below, the intrinsic signal-to-noise ratio (SNR) (of the NQR signal to the random thermal noise, primarily Johnson noise from the detector coil) is inherently low, and much effort goes into designing effective RF pulse sequences and detector coil geometries that maximize the SNR per unit time. To the degree that the noise is completely random, the improvement in SNR increases with the square root of scan time. A major complication in landmine detection is that radio frequency interference (RFI) from far field sources, such as AM radio transmitters, and near field sources, such as automobile ignitions and computers, creates substantial coherent noise that can be within the frequency regime of interest. Note (Figure K.2) that TNT frequencies are below 1 MHz, right in the AM band. Much effort has been devoted to reducing this problem—by coil design and by monitoring the RFI with a separate antenna and then subtracting the unwanted RFI signal. Nonetheless, the inherently weak NQR signals and the possible contamination by RFI dictate that *for present technology* it is not expected that the NQR detector will be satisfactory as a *primary* sensor for landmine detection in military applications. However, the exquisite selectivity suggests NQR is tailor-made as a *confirmation* sensor: e.g., when a primary sensor, such as an electromagnetic induction coil or ground-penetrating radar, indicates an anomaly, the NQR detector can be used as a confirmation tool to distinguish false alarms from landmines, without the need to mechanically probe the ground.

## STATE OF DEVELOPMENT

In the United States there are presently two programs to develop prototype NQR landmine detectors, both executed by Quantum Magnetics (QM). The U.S. Army Mine Countermine Division at Fort Belvoir, Va., sponsors development of a vehicle-mounted prototype NQR detector, designed initially as a confirmation sensor to clear mines from roadways. Under support from the Office of Naval Research and the U.S. Marine Corps, QM is also developing a prototype handheld landmine confirmation detector for the U.S. Marine Corps Systems Command, Quantico, Va.

## CURRENT CAPABILITIES

The prototype developments, above, represent the current capabilities. Both are for military applications, and development of an NQR detector for humanitarian purposes would use the basic technology but would take a somewhat different path—see below.

It is premature to report receiver operating characteristic curves.

## KNOWN LIMITATIONS

Not all explosives exhibit an NQR signal. In particular, some landmines, such as the PFM-1, employ liquid explosives that are not expected to be detectable [4]. Of course, NQR requires that the RF field must penetrate to the explosive, and so no NQR signal is obtained from a metal-encased mine. However, metal mines are easily found by metal detectors, and indeed the NQR detector acts as a crude metal detector itself. If the RF flux is excluded from a conducting volume, the effective inductance of the NQR detector coil is reduced and the presence of the metal case is indicated by an inability to resonantly tune the coil within the range of the variable tuning capacitor.

Both the NQR frequency and the relevant NQR relaxation times $T_1$ and $T_2$ are functions of temperature. The relaxation time $T_1$ determines how rapidly the pulse sequence can be repeated, and $T_2$ restricts the maximum length of the "spin echo" sequence used for TNT and tetryl detection. Roughly speaking, an acquisition of length $T_2$ can be obtained every $T_1$. Because the exact temperature of the mine is not known, one uses pulse sequence parameters that are broadly effective over a band of temperature. With some improvements, it is expected that estimating the approximate mine temperature to within 10–20°C should be adequate.

Detectability of RDX and tetryl is rather good by NQR, but for TNT the NQR relaxation times are less favorable, and the possible presence [5] of two crystalline polymorphs (monoclinic and orthorhombic) lead to weaker TNT signals. Finding small (50 g) antipersonnel TNT mines by NQR will be difficult, but is not ruled out. RDX and tetryl mines are much easier to find.

Most NQR detector coils use variants of a simple circular surface coil, for which the magnetic field intensity drops by about a factor of three at a distance of one coil radius along the coil axis. Beyond that distance, the field drops off more rapidly with distance, and so the detector coil radius determines the *approximate* useful depth of interrogation. A larger coil is an alternative, but more RF power is required for the transmitter, and because the "filling factor" (the volume fraction of the explosive divided by the effective volume irradiated by the coil) is reduced, the overall SNR can be reduced. Correspondingly, a significant increase in the "standoff distance" of the coil above the ground will also reduce the SNR of the mine.

In general, soil characteristics do not play a significant role in NQR detection. At these NQR frequencies (1–5 MHz), the RF field is not significantly attenuated, even for rather wet soils. The decrease in the effective *Q* of the detector coil corresponding to this RF loss results in a rather insignificant deterioration in SNR.

## POTENTIAL FOR IMPROVEMENT

Compared with combat engineering applications, there are some advantages to NQR for humanitarian demining. It is anticipated that increased scan time can be tolerated for humanitarian applications, especially if NQR is used as a confirmatory tool. For example, for TNT with a nominal $T_1$ relaxation time of 6 seconds at 20°C, two data acquisitions can be taken with a 6-second scan (using a delay of $T_1$ between acquisitions), but 51 acquisitions can be obtained in a 5-minute scan, giving an improvement of 5x in SNR. In other words, a 50-g mine would give the same SNR or detectability in 5 minutes as a 250-g mine in 6 seconds—an easy task for NQR. And, a 5-minute NQR scan sounds more appealing than digging by hand for 5 minutes.

Improvements in technology, especially the RF transmitter, and advanced NQR techniques such as "stochastic NQR"[6], should reduce the weight and the power requirements of the present military prototype detector and will be advantageous for humanitarian applications.

Other approaches that may be viable for humanitarian demining include combined NQR-NMR methods. For these a weak polarizing magnetic field is used, on the order of 10–100 G. Variants allow a cross relaxation between the nitrogen transitions and hydrogen NMR transitions, to either decrease the nitrogen $T_1$ relaxation time or to detect the nitrogen NQR transition as a perturbation on the much stronger proton NMR signal. These techniques have been explored to some extent in the laboratory for the past 50 years, and certain variants may be appropriate to humanitarian demining. See Nolte et al. [7] for a recent application, albeit one more suited to the laboratory.

Even with improvements, it is anticipated that NQR would still be integrated with some other detection method, such as electromagnetic metal detection.

## R&D PROGRAM TO REALIZE THE POTENTIAL

As indicated, NQR landmine detection for military applications is still in the prototype stage, and substantial improvement in the technology is expected as experience is gained. Extension to humanitarian demining can build on this experience, but significant effort should be devoted to adapt NQR to humanitarian demining.

The current five-year program to develop a prototype handheld NQR detector represents a projected investment of about $15 million for engineering and some science and technology (S&T). Further S&T work is necessary for the humanitarian approaches, perhaps $3 million worth. A program funded at about $1 million per year for research and development and $5 million per year for engineering development over a three-year period should make real progress toward an NQR detector specialized for humanitarian applications.

In the near term, the military version of the handheld NQR detector should be available in about four years. One might use those detectors for humanitarian purposes, assuming that there is no difficulty in the "dual use" of this technology. However, this appears to be at best a short-term solution because humanitarian demining requirements are different than those of the combat engineer.

## REFERENCES

1. A. N. Garroway, M. L. Buess, J. B. Miller, B. H. Suits, A. D. Hibbs, G. A. Barrall, R. Matthews, and L. J. Burnett, "Remote Sensing by Nuclear Quadrupole Resonance (NQR)," *IEEE Transactions on Geoscience and Remote Sensing*, No. 39, 2001, pp. 1108–1118.

2. C. Cumming, C. Aker, M. Fisher, M. Fox, M. la Grone, D. Reust, M. Rockley, T. Swager, E. Towers, and V. Williams, "Using Novel Fluorescent Polymers as Sensory Materials for Above-Ground Sensing of Chemical Signature Compounds Emanating from Buried Landmines," *IEEE Transactions on Geoscience and Remote Sensing*, No. 39, 2001, pp. 1119–1128.

3. J. A. S. Smith, "Nuclear Quadrupole Resonance Spectroscopy, General Principles," *J. Chemical Education*, No. 48, 1971, pp. 39–49; Y. K. Lee, "Spin-1 Nuclear Quadrupole Resonance Theory with Comparisons to Nuclear Magnetic Resonance," *Concepts in Magnetic Resonance*, No. 14, 2002, pp. 155–171.

4. *Jane's Mines and Mine Clearance*, 2001–2002.

5. G. R. Miller and A. N. Garroway, *A Review of the Crystal Structures of Common Explosives Part I: RDX, HMX, TNT, PETN, and Tetryl*, NRL Memorandum Report NRL/MR/6120-01-8585, October 15, 2001.

6. A. N. Garroway, J. B. Miller, D. B. Zax, and M-Y. Liao, "A Means for Detecting Explosives and Narcotics by Stochastic Nuclear Quadrupole Resonance (NQR)," U.S. Patent 5,608,321, March 4, 1997.

7. M. Nolte, A. Privalov, J. Altmann, V. Anferov, and F. Fujara, "$^{1}$H–$^{14}$N Cross-Relaxation in Trinitrotoluene: A Step Toward Improved Landmine Detection," *J. Phys. D: Appl. Phys.*, No. 35, 2002, pp. 939–942.

Appendix L

# X-RAY BACKSCATTER (PAPER I)

*Lee Grodzins, American Science and Engineering*[1]

## OVERVIEW

Most of the concepts presented in this paper feature AS&E's patented backscatter technology, which uses a scanning pencil beam of x rays. Backscatter technology should be most applicable as a tool for identifying antipersonnel mines that are planted no more than 3 inches below the surface or that are lying on the surface but are hidden by camouflage or vegetation. Many groups have researched standard x-ray backscatter over the past years as a potential mine detection tool. That success has been limited at best. We present here concepts that, as far as we know, have not been implemented, although our experience with similar systems indicates that it would be relatively straightforward to do so. We present these ideas, all studied by Monte Carlo simulations and most of them patented, as new techniques that have the potential for eliminating some of the problems encountered in the past using x-ray systems for humanitarian demining.

---

[1]American Science and Engineering Inc. (AS&E) is pleased to submit this appendix to RAND for use by its Mine Detection Technology Task Force. As a result of a request by RAND to Dr. Lee Grodzins, MIT Professor Emeritus, and Vice President, Advanced Development, AS&E, we have assembled a series of concepts that could be employed to detect a wide variety of landmines in a relatively short period and have the potential for being developed into deployable systems.

## NATURE OF THE PROBLEM

According to reported estimates, approximately 100 million antipersonnel mines now lie hidden in the ground in 64 countries worldwide. These mines remain active for many years and kill or maim 25,000 civilians every year. The task of finding and removing these mines is monumental. Many of the mines that were close to or on the surface have become more deeply buried or are heavily overgrown with vegetation. In addition, many of the newer mines are clad with plastic, rather than metal, and the use of metal detectors to locate these mines is no longer effective. There is therefore an urgent need for new technologies to be developed that can effectively define an area containing antipersonnel mines, with minimal false alarms. Figure L.1 depicts the type of antipersonnel mine that drives both the mapping and detection requirements.

Although there has been significant effort aimed at the detection and clearing of various kinds of mines, what is needed is a methodology and system implementation focused on the safe and rapid delineation of boundaries of areas containing mines. This objective can be

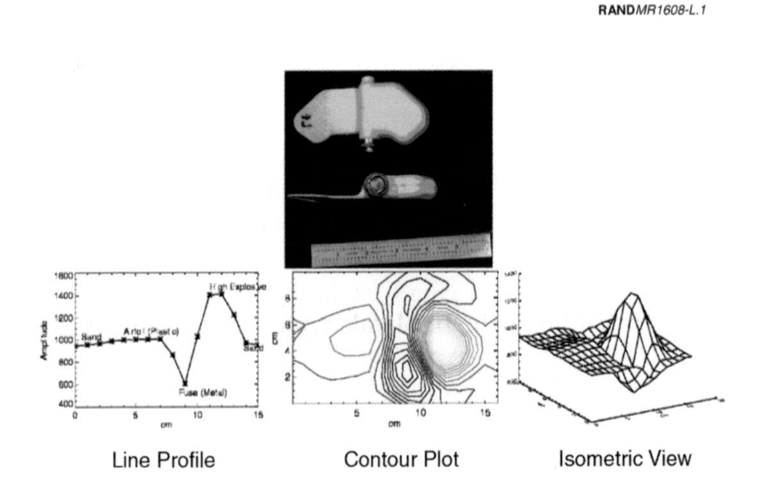


**Figure L.1—High-Resolution Image of a "Butterfly" Antipersonnel Landmine**

viewed from both a military/humanitarian and a commercial/ industrial standpoint. In a military/humanitarian sense, governments seek to ensure that civilians relocating to areas where there have been conflicts will not be in danger of mutilation or death from contact with antipersonnel mines. Likewise, as commercial ventures begin to reclaim and reuse land as part of the Base Realignment and Closure process, there are usually areas of ordnance that need to be identified, mapped, and eventually cleared. The system that performs the detection and mapping operation must cover a large area—on the order of thousands of square feet—quickly, reliably, and safely.

## AS&E'S POTENTIAL SOLUTION SET

The following paragraphs describe a number of potential approaches using x-ray backscatter to address the demining problem.

### Background and General Remarks[2]

The signal of x rays scattered by a buried object is proportional to the product of a number of factors, which are as follows:

- The intensity of the x rays in the pencil beam emitted by the x-ray source.
- The attenuation of the x rays before the scatter.
- The probability of scattering in the back direction. (This probability depends inversely on the absorption power of the material to the incident and to the backscattered x rays. Organic materials typically absorb only a small fraction of the x rays, so that the scatter probability is high. Metals typically are strongly absorbing, and the scatter probability is low. Thus, organic materials are bright and metallic objects are dark in the image.)

---

[2]This appendix contains minimum background and detail. The background material for x-ray backscatter is well summarized in P. Horowitz, K. Case, et al., *New Technological Approaches to Humanitarian Demining,* McLean, Va.: MITRE, JSR-96-115, 1996.

- The attenuation of the x rays as they pass back out of the object and traverse the soil on the way to the detector.
- A factor that depends on the product of the density (ρ) and the thickness (t) of the object along the x-ray path. For thick objects, the dependence is weak. If the thickness is not known, then one is unable to get any measure of the density.
- The solid angle and efficiency of the backscatter detectors.

Single scattering dominates for objects on or just under the surface, but multiple scattering increasingly dominates as the sought-for object is more deeply buried. Explosive material, soil, and organics preferentially scatter, rather than absorb, the higher-energy x rays that are required to see objects below the surface. In practice, multiple scattering as well as single scattering must be considered to make an effective detector.

The most important information is the image itself. The operator must see a high-quality image to identify the objects being viewed. AS&E images typically have about 2-mm-diameter resolutions, adequate for even small landmines. At 1 ms per point, the area scan rate can be 3 sq ft per minute. The resolution in AS&E backscatter systems can be varied in situ. At a resolution of 4 mm × 4 mm, probably adequate for many purposes, the scan rate can be more than a square meter per minute.

We note that all the devices described in this appendix can be mounted on a remote-controlled vehicle, with only the detectors suspended over the minefield by a boom. To produce an undistorted image the speed of the vehicle would need to be either regulated or accounted for, as AS&E does for its "X-Ray Van: Drive-By Backscatter Imaging" project.

Figure L.2 shows the ability of an AS&E backscatter image to clearly indicate the shapes of solid objects inside a suitcase despite a cluttered background. (We note that the monitor image is considerably crisper and more photographic-looking than the printed image in Figure L.2.)

To inspect terrain for antipersonnel mines, the system would be oriented facing the ground. The backscatter system might be applied

RANDMR1608-L.2

NOTE: An example demonstrating the ability of backscattered x rays to distinguish solid objects against a cluttered background. Unlike transmission x-ray images, the backscatter image is formed without the need to place an imaging screen or other hardware on the far side of the subject.

**Figure L.2—AS&E Backscatter Image**

as the second stage of a two-stage process in which the first stage identifies suspect objects, and backscatter imaging verifies each threat. For example, the AS&E backscatter system could be integrated with the water-jet drilling system developed by the University of Missouri–Rolla as shown in Figure L.3. The water-jet system has been shown to have the potential for locating buried antipersonnel mines without detonating them, but it is unable to distinguish between mines and a variety of other solid objects. The use of backscatter imaging of all suspect objects could be a method to quickly dismiss false alarms and to help to identify the type of mines discovered prior to excavation.

The backscatter system in Figure L.3 employs a hollow spoked rotating wheel to create a scanning pencil beam of x rays. AS&E is developing a CRT (cathode-ray tube)–based approach to x-ray scanning to reduce weight and significantly improve flexibility. Figure L.4 shows that approach to x-ray beam generation combined with water-jet clearing of debris and topsoil.

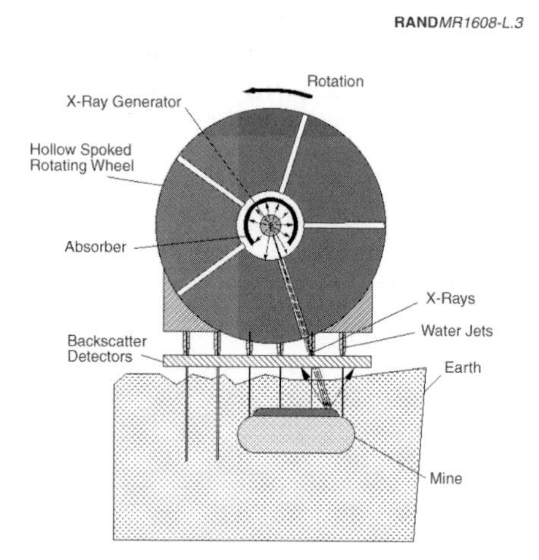


**Figure L.3—Illustration of the Standard AS&E Backscatter System Adapted to Work in Conjunction with a Water-Jet Hole Drilling System**

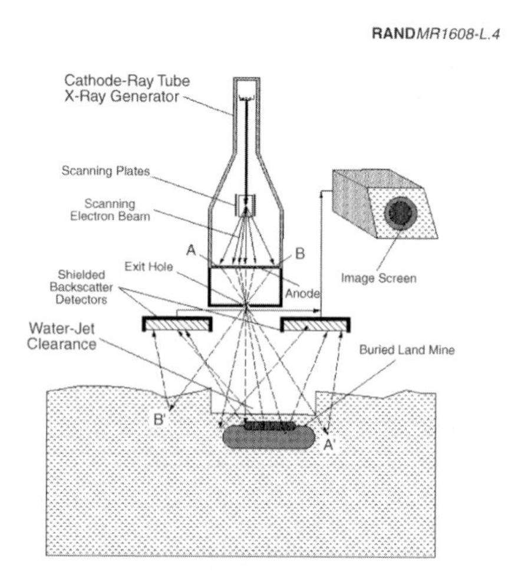


NOTE: This system works in combination with a water-jet system to clear surface soil from above the suspect object to allow imaging at greater depths than are normally possible with backscattered x rays.

**Figure L.4—Illustration of a Scanning Electron Beam Backscatter System**

In some terrain, it may be practical to carry out airborne searches with the x-ray mine detection system suspended from a helicopter, but such a modality faces many problems not encountered with a ground-based system.

## Shallow Angle Compton Scatter Detection

One method to increase ground penetration is shown in Figure L.5. The incident radiation enters the ground at a shallow angle, and the detectors are positioned to receive radiation that has scattered off the target at an angle in the 90° range. The energy of this scattered x ray is significantly greater than that of the 180° backscattered x ray, potentially allowing penetration depths greater than 5 cm.

## Focused-Depth Compton Backscatter Imaging

The problem of imaging objects buried just below the surface is complicated by the irregular distribution of material (uneven terrain, rocks, debris, surface vegetation, etc.) on the surface. These features, collectively known as "surface clutter," contribute strongly to the background counts and reduce the signal-to-noise ratio of the land-mine image. AS&E has pioneered collimation methods to reduce the

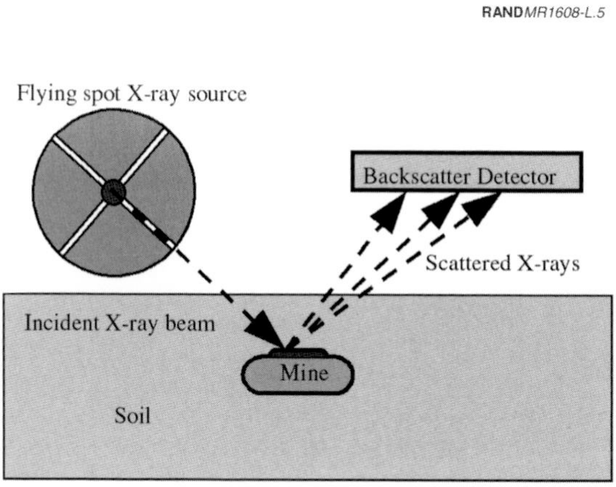

RANDMR1608-L.5

**Figure L.5—Backscatter System Modified to Use Forward Scatter Radiation**

signal from near-field clutter by adding collimation vanes in front of the backscatter detectors, as shown in Figure L.6. These vanes block photons emanating from depths other than the depth of interest. The collection efficiency of this geometry as a function of depth is shown in Figure L.7. Our experiments show that the signal intensity changes by a factor of two over a distance of just 1.2 cm.

## Effective Atomic Number and Density Measurement with Compton Backscatter Imaging

Backscatter imaging is an effective modality to distinguish mines from false positive alarms (e.g., rocks, scrap metal, roots) by measuring the effective atomic number of the detected objects. The intensity of single backscattering with incident energies above about 100 keV is quite sensitive to large differences in the atomic number of the scatterer, such as plastic and steel. Single backscatter is, therefore, quite capable of distinguishing a chunk of metal from a plastic mine. However, to distinguish a plastic mine, with its effective Z of about 7, from rocks, with an effective Z of about 14, requires that one use

RAND*MR1608-L.6*

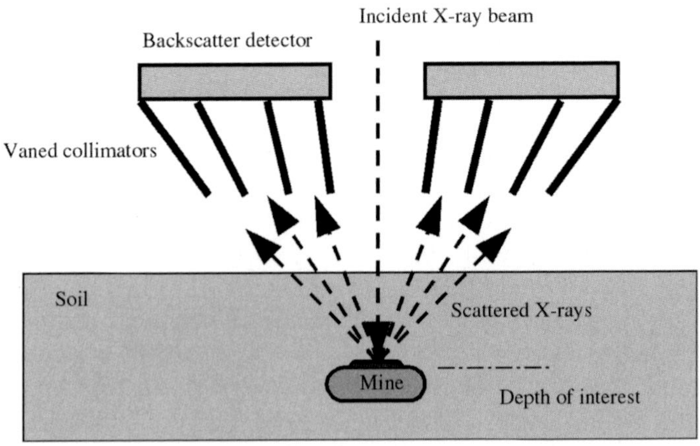

**Figure L.6—Schematic Diagram of Vaned Collimator Imaging Technique**

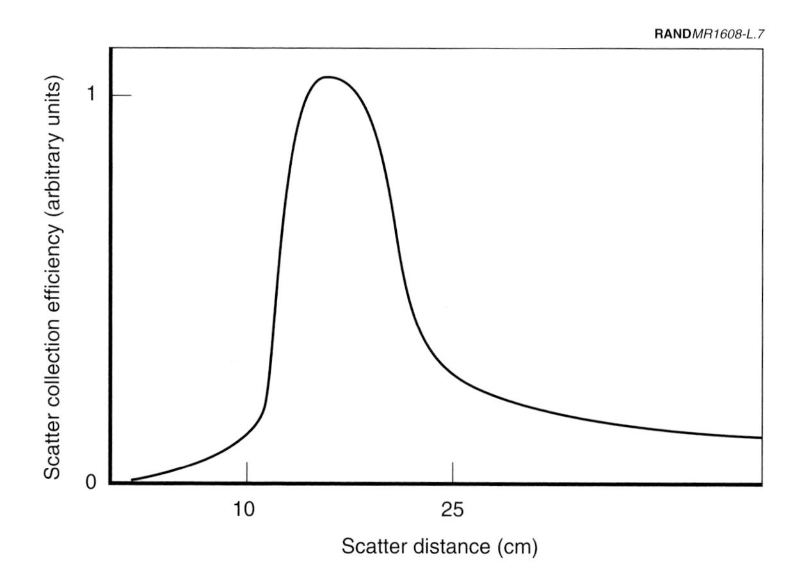


**Figure L.7—Collection Efficiency as a Function of Depth in Inspection Tunnel Using Vaned Collimators**

much lower x-ray energies (hence much reduced penetration) or a more sensitive backscatter technique. Calculations show that backscatter radiation intensity resulting from multiple scattering in the object is more sensitive than single scattering to the atomic number of the object.

Figure L.8 shows collimators that block photons that have a direct line of flight from the point of interaction in the target to the detector, while passing photons that have moved some lateral distance in the target before leaving the target. Other groups have used this method, first developed in the 1970s by F. Roder.

AS&E's method would extend the one-detector technique of others to the use of several independent concentric rings of detectors. Computer simulations indicate that the ratio of responses in the detectors gives a measure of the object's density, while the sum of signals in the set of detectors gives a measure of the effective atomic number.

RANDMR1608-L.8

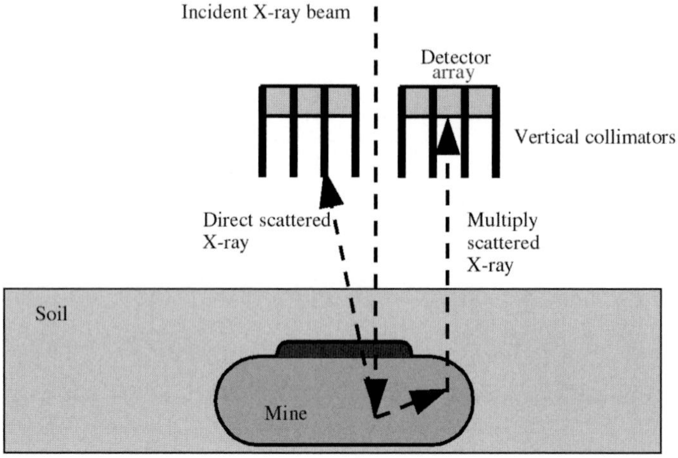

Incident X-ray beam
Detector array
Vertical collimators
Direct scattered X-ray
Multiply scattered X-ray
Soil
Mine

**Figure L.8—Schematic Diagram of Backscatter Detector and Collimator Configuration to Measure Atomic Number**

## Backscatter System Configured for Optimum Density and Effective Atomic Number Measurements

Effective atomic number alone is often not an adequate metric to identify landmines because nonmetallic mines have a similar effective atomic number to that of organic material. The addition of density information can considerably improve our ability to separate either metal or plastic mines from a background including not only metallic objects and rocks (silicates) but also roots and other organic debris.

Figure L.9 is a schematic of a detector system that measures both the effective atomic number and the density. The inner set of detectors counts multiply scattered x rays, providing atomic number information. The outer set of detectors are collimated to restrict the thickness of the measuring volume and hence obtain density information. It is important to emphasize that, because the position of the x-ray beam relative to the ground is the same for all detected counts, one can

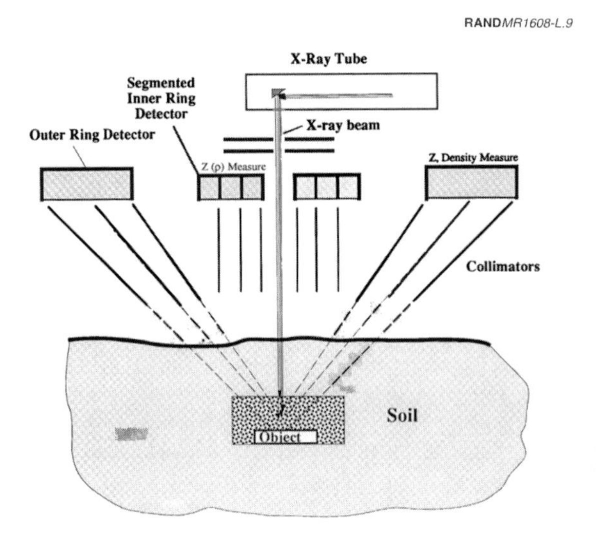


**Figure L.9—Schematic Diagram of Backscatter Detector and Collimator Configuration to Simultaneously Measure Atomic Number and Density**

sum the detector counts to produce a crisp image. If sufficient counts are available in individual detectors, then the images of each can give additional information. AS&E has carried out extensive and effective measurements of density distributions in luggage with the collimated detectors at right angles to the x-ray beam (AS&E's patented Side-Scatter Tomography). A right-angle measurement is not practical for landmine detectors, so the collimation cannot be as effective as it is for luggage. However, a depth interval can be defined well enough so that in many situations the deduced density can discriminate explosive material with densities in the 1.4–1.7 g per cubic centimeter range from such materials as roots that have densities less than 1 g per cubic centimeter.

## Underground Source Transmission Imaging

All the systems described above use backscatter imaging in part because the problem of imaging objects in the ground is perceived to be a necessarily one-sided imaging task. A two-sided imaging solution, such as transmission x rays, can be made possible if one can place a source of intense radiation in a hole drilled to some distance below the surface. Figure L.10 outlines this concept.

The hole is drilled in a safe area adjacent to the area to be investigated. A transmission x-ray source or an x-ray generator is inserted in the hole so that radiation is transmitted through the earth and any buried objects are imaged at a detector similar to the digital-imaging detectors used for medical applications. For shallow mines, nearly the entire area below the detectors can be inspected. For deeper mines, a smaller area of ground will be within the volume that is penetrated by x rays. If mines are detected, their positions are noted and then used to direct demining work. Once the area is determined to be safe, the equipment is broken down and a new hole or holes can be drilled in the area that has been demined to reach farther into the minefield.

A small system using holes on the order of 20 cm deep could be used to identify potentially dangerous objects located by other mine detecting methods such as water-jet drilling and acoustic sounding.

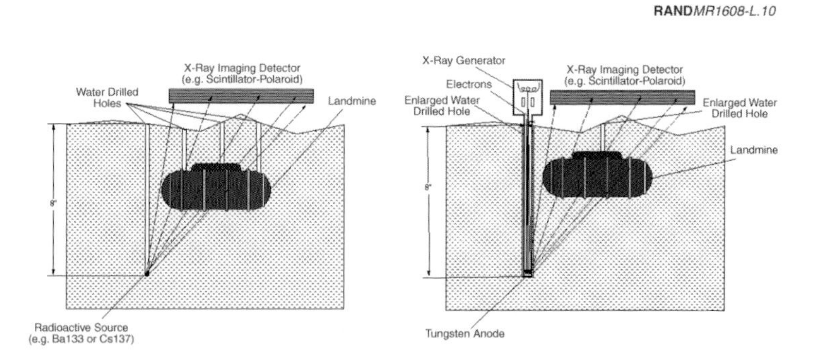


Left: Illustration of x-ray transmission imaging with the radioactive source placed at the bottom of a water-jet drilled hole. Right: Illustration of x-ray imaging with a thin-tube x-ray generator inserted into an enlarged water-jet drilled hole.

**Figure L.10—Two Illustrations of X-Ray Imaging**

A larger system using deeper sources might be used as the primary method of locating mines.

The use of a radioactive source raises the question of the environmental risk in the event that an accidental detonation destroys the equipment and scatters fragments of the source. Using a gaseous radioactive source, such as the long-lived krypton-85, which emits 0.514 MeV gamma rays, could mitigate this hazard. In the event of damage to the source, the krypton gas would dissipate, leaving no contamination of the surrounding soil.

## TYPICAL PROGRAM PLAN

To investigate the potential techniques described in the previous section, AS&E would follow a program task outline as described below, with a nominal schedule as shown in Figure L.11.

Tasks include the following:

1. **Problem Definition:** What specific type of mine needs to be detected in this program? What are its characteristics? How is it deployed? What is the nature of the surrounding area—soil, rock, etc.?

RANDMR1608-L.11

Months After Contract Award

| Task | Mo. 1 | Mo. 2 | Mo. 3 | Mo. 4 | Mo. 5 | Mo. 6 | Mo. 7 | Mo. 8 | Mo. 9 | Mo. 10 | Mo. 11 | Mo. 12 |
|---|---|---|---|---|---|---|---|---|---|---|---|---|
| Problem Definition | | | | | | | | | | | | |
| Potential Solution Set Definition | | | | | | | | | | | | |
| Technology Analyses | | | | | | | | | | | | |
| Selected Technology Demonstration | | | | | | | | | | | | |
| Analyses/Evaluations | | | | | | | | | | | | |
| Prototype Development | | | | | | | | | | | | |

1 2 3 4 5 6 7 8 9 10 11 12 13 14 15 16 17 18 19 20 21 22 23 24 25 26 27 28 29 30 31 32 33 34 35 36 37 38 39 40 41 42 43 44 45 46 47 48 49 50 51 52

**Figure L.11—Nominal Program Schedule**

2. **Potential Solution Set Definition:** How could this problem be solved? Identify the methods under consideration and briefly describe them.
3. **Technology Analysis:** Identify the advantages and disadvantages of each technique under consideration. Qualities should include: technology maturity, projected operational concept, type of mine detected, known or suspected limitations, speed of operation, false alarm rate, projected system price, and life-cycle cost.
4. **Selected Technology Demonstration:** Design and implement a quantitative and qualitative test program that will provide representative data about the selected technology.
5. **Analyses:** Analyze the results of the demonstration and provide a report containing results and recommendations for future work.
6. **Prototype Development:** Design, fabricate, and test a system that incorporates the selected technology.

Appendix M

# X-RAY BACKSCATTER (PAPER II)

*Alan Jacobs and Edward Dugan, University of Florida*[1]

## SYNOPSIS

In the presence of realistic soil inhomogeneities and surface clutter, employment of high-energy photon radiation (x rays or gamma rays) for landmine detection requires mine-feature imaging with centimeter spatial resolution. An x-ray backscatter method has been developed with photon detection efficiency sufficiently high such that an electric power requirement of only hundreds of watts yields a soil surface interrogation rate of 1 sq m per minute. Extensive laboratory measurements, and one test on the mine lanes at Fort A. P. Hill, Va., have demonstrated such results that it is estimated that mine detection probability is unity, false alarm probability is 0.03, and false alarm rate is 0.1 per square meter for mine depths of burial up to 5 cm. These results were achieved with a first developmental system, the components of which can be easily improved, and the imaging protocol of which can be modified to yield the same, or better, performance parameters up to mine depths of burial of 10 cm or more. This improved system will have weight less than 100 kg, will have about 1-m dimensions, and will be sufficiently rugged for real field applications. A manufactured unit cost will be on the order of $10,000.

---

[1]This paper is based on research conducted at the University of Florida, under the sponsorship of the U.S. Army.

## BACKGROUND

Conventional x-ray radiography employs the transmission of photons through an illuminated object to produce an image. The image is formed by detection of the penetrating, uninteracted x-ray field and depends on the geometrically projected attenuation properties of the internal structure of the object. Clearly, inability to access the penetrating field renders unthinkable the use of conventional radiography for examination of objects, such as landmines, buried in soil. Compton backscatter imaging (CBI) is an x-ray radiography technique that utilizes detection of photons scattered by the internal contents of an object to form images. In typical landmine detection situations, a significant fraction of the illumination photons are scattered by the mine-soil interface and emerge through the soil surface traveling toward locations above the surface where detection is possible. This is the physical basis of CBI and forms the foundation of the results and ideas expressed in this appendix.

The CBI approach is not new. The first published account by Odeblad and Norhagen [1] describes a measurement system using a collimated gamma-ray source and a collimated scintillation detector. The work measured relative electron densities in object internal volumes formed by the intersection of the fields of view of the detector and source collimators. There are many subsequent published investigations wherein the general idea of a localized illumination viewed by a localizing detector aperture is employed to accomplish CBI. A notable commercial device, the ComScan system, has been applied to image aircraft structures as well as buried landmines. In both applications, the lengthy time required to acquire an image renders common usage impossible.

Another, very relevant CBI approach (discussed below) by Towe and Jacobs [2] uses a collimated x-ray source and a small, but uncollimated, detector employed to sense large-angle (ca. 180°) backscattered photons. Energy modulation of the x-ray generator is used to produce two images. Subtraction of the lower-energy image from the higher-energy image yields a tomographic representation of a layer within the object, which is transverse to the illumination beam direction. Significant surface structure of an object can thereby be removed from a CBI image without severely limiting the fraction of the viewed emerging photon field.

Prior to 1975, a number of attempts were made to use backscattered photons (either x rays or gamma rays) to detect buried nonmetallic mines. A summary of these efforts is reported by Roder and Van Konyenburg [3]. To achieve significant detection efficiencies, it was required to apply all the measurement systems to highly idealized situations. Specifically, the soil was required to be homogeneous with a plane surface. Realistically expected inhomogeneities, such as roots and rocks, and surface structure, such as potholes and tire tracks limited the mine detection efficiency.

In 1986, Fort Belvoir personnel proposed to the University of Florida (UF) that the x-ray generator energy modulation variant of CBI be applied to the landmine detection problem. It was clear early in the project that the method was not directly applicable to mine detection if desired speeds of image acquisitions were to be achieved. Data acquisition rates for either military or humanitarian landmine detection would require detection efficiency for relevant, information-bearing photons that is orders of magnitude higher than all previously developed CBI techniques, including the uncollimated-detector, energy-modulation approach with its relatively high detection efficiency. In response to this dilemma, UF developed a totally new CBI approach, called lateral migration radiography (LMR). In the remainder of this appendix, the physical concepts of the LMR technique are briefly outlined; the remarkable mine signatures (images) obtained for actual mines, buried in idealized laboratory situations, are reviewed; results obtained during three days of trials with a first-generation mobile LMR system on the mine lanes at Fort A. P. Hill are summarized; and development ideas for a practical embodiment of the method, with expected improved performance, are suggested.

## LATERAL MIGRATION RADIOGRAPHY

All "conventional" CBI systems rely on the selective detection of photons that have scattered from only one object to form an image. Object surface irregularities and internal inhomogeneities, as well as the undesired detection of multiple-scatter photons, obstruct and corrupt such first-scatter dependent techniques. Highly localizing collimators on both x-ray generator and scatter-field detectors are required to extract useful subsurface structure information. This leads to high source strength and slow imaging system operation.

The technique of LMR is a new imaging modality that employs both single-scattered photons and the lateral transport of multiple-scattered photons to form separate images. A summary of the method is published by Su et al. [4]. Very large area scintillation detectors significantly reduce the required x-ray source strength and image acquisition time. The present UF LMR systems use two types of detectors to form images. Uncollimated detectors sense predominately once-scattered photons and primarily generate images of surface and near-surface features. Properly positioned and collimated detectors sense predominately multiple-scattered photons. The contrast in the collimated detector images is primarily due to the photon lateral transport in the object, which is sensitive to both electron density and atomic number variation of the object medium along such transport paths. The LMR configuration allows CBI (with high photon collection efficiency) of objects that contain extended electron density or atomic number discontinuities in the paths transverse to the incident illumination direction. The multiple-scattered photon distribution is influenced by the first-scatter distribution and thereby is subject to object surface variation. The separate sensing of the first-scatter photons allows for effective removal of the surface-influenced component of the collimated detector image by subtraction.

## LABORATORY SYSTEM AND RESULTS

The UF laboratory landmine imaging system includes a pair of uncollimated detectors (each with a sensitive area of 300 sq cm) and another pair of detectors (each with an area of 900 sq cm), collimated by lead sheets against sensing once-scattered x rays. The particular configuration employed to acquire the images presented herein is illustrated in Figure M.1. To generate an image, the x-ray illumination beam should raster in the gap between the two uncollimated detectors and also move with the detectors in a direction orthogonal to this raster. However, in the existing laboratory image acquisition system, the illumination beam remains stationary, and a large soil box, in which mines are buried, moves in the two orthogonal directions. This type of soil illumination is necessitated in the laboratory image acquisition system because of constraints of the available (cumbersome) x-ray generator. In contrast, Figure M.2 shows the mobile LMR mine detection system used in the tests at Fort A. P. Hill. The required orthogonal scan motions of the x-ray illumination beam are provided by a rotating source collimator and a linear motion of the mounting platform.

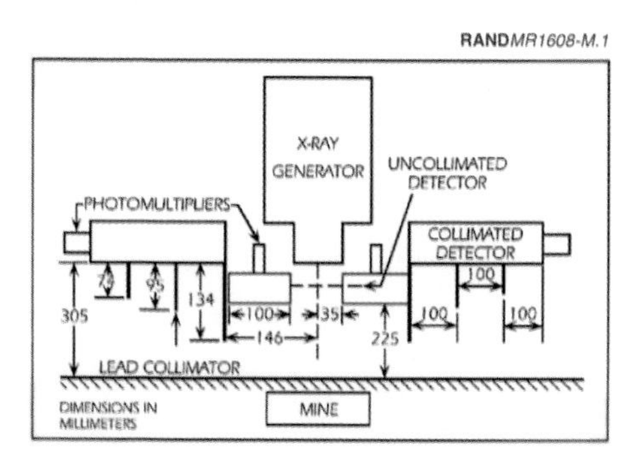


**Figure M.1—Configuration Used to Acquire Images**

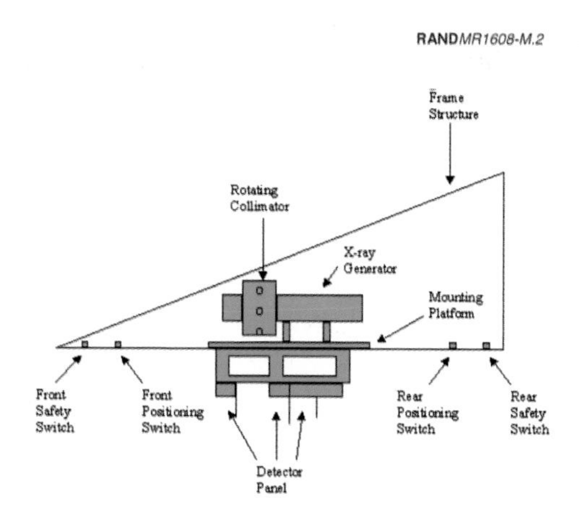


**Figure M.2—Drawing of Field Test LMR Mine Detection System (XMIS)**

The results of LMR imaging of three plastic buried landmines (M-19 antitank, TMA-4 antitank, and TS/50 antipersonnel) are included herein as Figures M.3–M.6. These images are part of the output of laboratory measurements in which LMR was used to image 12 types of actual landmines provided by the U.S. Army. The acquired images demonstrate that detection is possible with burial depths ranging from the soil surface to 10 cm. Moreover, the images (signatures) are so definitive that, under the idealized laboratory conditions, clear identification of mine type can be accomplished. When combined with the exterior mine shape, interior air volumes offer unique signatures. In the laboratory environment, the LMR technique, for near-surface buried mines, seems to be free from the problem of false positive alarms.

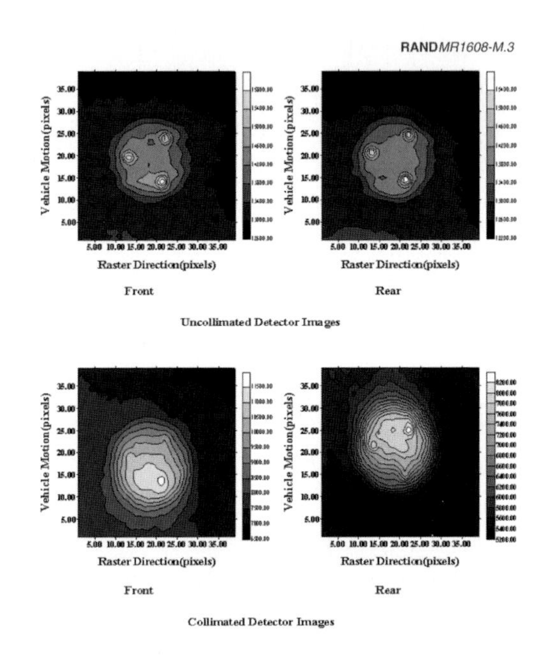


NOTE: These are LMR images acquired in each of the four detector array components of a TMA-4 antitank mine with 2.5-cm depth of burial using 1.5-cm pixel size. Note the three fuse-well details in the uncollimated images and the 10-pixel offset between collimated front and rear detector images—a direct measure of the depth of burial of the mine. Maximum/minimum image intensity ratios: uncollimated = 1.29, collimated = 1.73.

**Figure M.3—LMR Images of TMA-4 Antitank Mine**

In each of Figures M.3–M.6, the caption includes LMR imaging parameters, burial conditions, and salient image signature features. It should be emphasized that the crucial, high-intensity LMR regions in these figures are generated by the mine-interior air volumes, not the mine surface, and that these intense signatures are certainly identifiers of the presence of a buried landmine if not, in some cases, a unique response of the mine type.

Clearly, the image results typified by Figures M.3–M.6 are attributed to the laboratory-imposed conditions of a homogeneous soil with a level plane surface. Numerous measurements have been made with the laboratory system using various soil surface structure and irregularities. In all cases, image subtraction yields mine images of quality

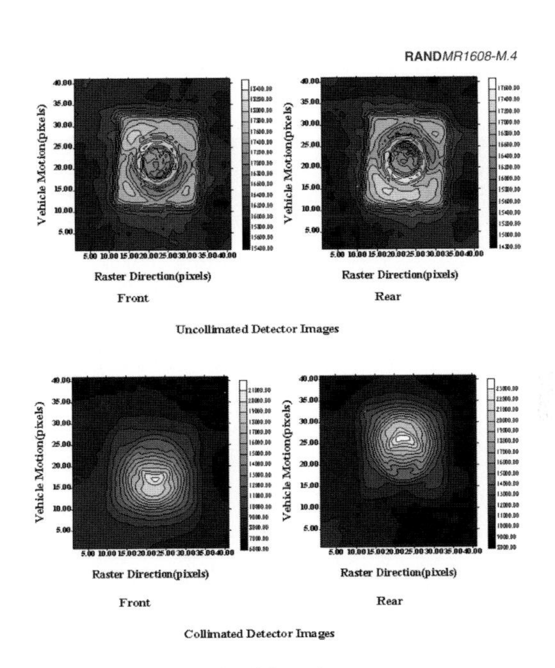


NOTE: LMR images acquired in each of four detector array components of an M-19 antitank mine with 2.5-cm depth of burial using 1.5-cm pixel size. Note the characteristic square shape of the plastic casing and the details of the cylindrical fuse well in the uncollimated images and the depth-of-burial-dependent offset of the collimated images. Maximum/minimum image intensity ratios: uncollimated = 1.21, collimated = 3.33.

**Figure M.4—LMR Images of M-19 Antitank Mine**

similar to those shown here. It is also clear from these measurements that LMR detection has an upper limit of mine depth of burial near 10 cm. The real test of this technique is its applicability in a field environment. One set of tests has been accomplished in the mine lanes at Fort A. P. Hill.

## FIELD SYSTEM AND RESULTS

The characteristic of the LMR mine image acquisition process that leads to efficient use of input electric energy is that all photons emerging from the mine and soil can be employed in forming the image. This leads to the inclusion of large area detectors in a system design. In both the UF laboratory and field test versions, photon detection efficiencies are sufficiently high that good image quality

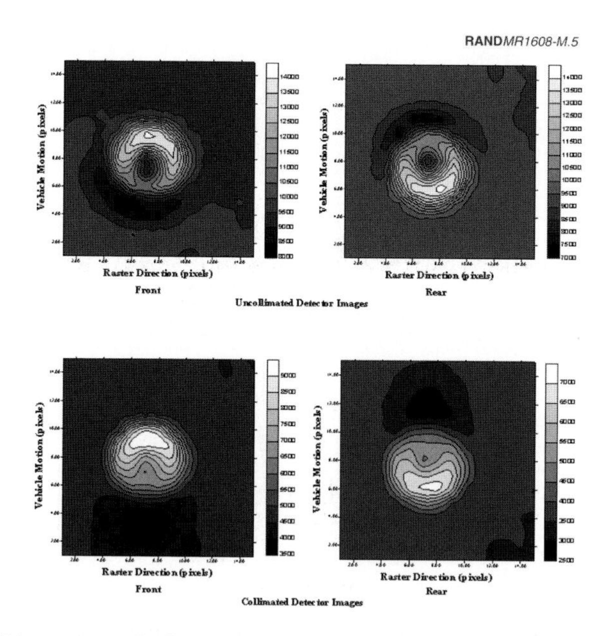


NOTE: LMR images acquired in each of four detector array components of a surface-laid TS/50 antipersonnel mine using 1.5 cm pixel size. Note the characteristic fuse-well details as well as the distinct "shadow" of the mine due to mine protrusion above the soil surface. Maximum/minimum image intensity ratios: uncollimated = 1.94, collimated = 2.86.

**Figure M.5—LMR Images of Surface-Laid TS/50 Antipersonnel Mine**

implies only about 2 million illumination x-ray photons per soil surface pixel. Based on the x-ray generators and geometric configurations chosen, this photon number translates into a calculated generator electric energy requirement of 1 joule per pixel. A pixel illumination (beam) size of 1.5 cm × 1.5 cm provides sufficient resolution for both antitank and antipersonnel mines. The calculated electric energy usage efficiency and a presumed soil surface interrogation rate of 1 sq m per minute implies an x-ray generator power requirement of about 200 watts and a pixel dwell time of about 10 milliseconds. The UF field test LMR landmine detection system, illustrated in Figure M.2, is designed based on the above presumptions. The system is identified as the "x-ray mine imaging system" (XMIS) and employs an air-cooled, commercial 160 kVp x-ray generator with focal spot positioned about 80 cm above the soil surface. The focal spot location is surrounded by a 10-slit rotating collimator

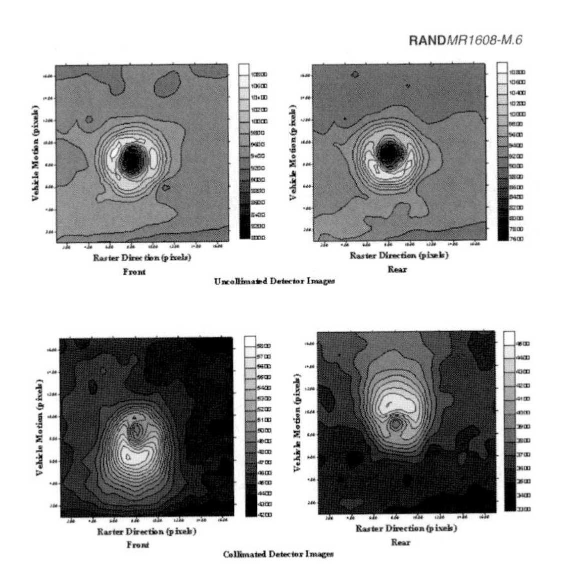


NOTE: LMR images acquired in each of four detector array components of a flush-with-surface TS/50 antipersonnel mine using 1.5 cm pixel size. Note the enhanced fuse-well details showing the small steel fuse springs, the absence of "shadow" characteristic of a surface landmine, and the 3-pixel depth-of-burial offset in the collimated images. Maximum/minimum images intensity ratio: uncollimated = 1.41, collimated = 1.42.

**Figure M.6—LMR Images of Flush-with-Surface TS/50 Antipersonnel Mine**

to provide one of the illuminating beam scan directions. For the desired imaging rate of 1 sq m per minute, the above design choices imply the easily achieved conditions: 6 rpm rotation of collimator; 1.5 cm per second linear scan direction motion; 1 second per single-line scan; and about 30 milliseconds dwell time per 1.5 cm × 1.5 cm pixel. The pixel illumination is contiguous in both scan directions, and there is a near-zero dead time in the data acquisition process. With reference to Figure M.2, the visible detector dimensions (widths) are 5 cm for the uncollimated and the small-collimated detectors and 20 cm for the large-collimated detector. The three-detector scintillation panels are 140 cm in dimension (length) perpendicular to the plane of the figure. The scintillation panels are 5 cm thick. The small- and large-width detector panels provide sensitive areas, which are 700 sq cm and 2,800 sq cm, respectively. The entire XMIS, as shown in Figure M.10, is approximately 150 kg including the x-ray generator controls and high-voltage power supply. As configured, and with the detector panel bottom surface 30 cm above the soil, XMIS provides a surface scan region of 50 cm × 50 cm (about 1,000 1.5-cm-sized pixels). As employed in a series of measurements on the mine lanes at Fort A. P. Hill, the x-ray generator electric power was set at about 700 watts and region image (frame) acquisition time was about 30 seconds. These values imply about an order of magnitude higher electric energy per pixel than the ideal calculated value. Some reasons for this are discussed below. The XMIS was supported by a moveable trailer (also used for transport from UF). The combined system, in operation, is shown later in Figure M.11.

As an aid in addressing the image processing employed in the field test, a review of the LMR images presented as Figures M.3–M.6 (where no processing is employed) is useful. Note that the extraordinary image detail in the uncollimated detector images is significantly blurred in the collimated images, but the collimated image "signal" is always more intense. The obtained uncollimated detector image detail is, with certainty, due to the ideal conditions of the laboratory. Surface structure and inhomogeneities do provide major image features in realistic situations. In fact, the reason why uncollimated detector images are useful when imaging buried mines is that cloaking of the mine image in the collimated detector images by soil surface features can be effectively removed by image subtraction. Moreover, note that in Figures M.3, M.4, and M.6 the center of intensity of the two collimated detector (front and rear) images is shifted (backward and forward, respectively). As implied in the figure captions, the magnitude of this shift can be employed to deduce the approximate depth of burial of the mine. Of greater importance here is that the two "views" of the mine and soil provide an image-processing scheme to selectively enhance the presence of a mine. Note that in Figures M.3–M.6, the rotating collimator-induced scan is termed "raster direction" and the linear motion-induced scan is termed "vehicle motion," which are certainly misleading designations for the XMIS embodiment of LMR but could be meaningful in larger-scale versions of mine detection systems.

The image processing sequence applied to the three-detector image set required in the field test is the following:

1. Intensity-normalize the image set.
2. Subtract the normalized uncollimated detector image from each of the two normalized collimated detector images.
3. Obtain the average of intensity of the image formed by the multiple of the resulting images (of step 2) shifted relative to each other in the linear scan (front/back) direction as a function of image shift.
4. Display the final multiplied image for the case of the maximum value obtained in step 3.

Figures M.7–M.9 are examples of the mine imaging results obtained in the Fort A. P. Hill test. Figure M.7 shows the case of a VS1.6 antitank mine (22 cm in diameter) at a 2.5-cm depth of burial in dirt. This result is included here as a comparison with the more relevant results in Figures M.8 and M.9, which are images of TS/50 antipersonnel mines (9 cm in diameter). Figure M.8 conditions are 1.3-cm depth of burial in featureless surface dirt. More relevant is the result in Figure M.9, which is for conditions of 5-cm depth of burial in dirt covered with a significant amount of natural foliage and rocks. In fact, note the displacement of the mine image from the image center. In this case, it was difficult to determine from surface markers where the mine was actually situated when the scan location was chosen.

The field test at Fort A. P. Hill was certainly inadequate. UF was invited to take images of sites where ground-penetrating radar methods had yielded consistent false positive alarms. Of the 30 such sites imaged, in only six cases did the XMIS images yield signatures with any mine-like features, and in only two of these did the processed image indicate a possible buried mine. The 12 buried mine cases interrogated were insufficient to glean convincing conclusions. However, for depth of burial less than 5 cm, the field test results are similar to those found for the cases shown herein. As discussed in the next section, XMIS should not be considered a prototype but rather only a demonstration system for the LMR approach to landmine detection.

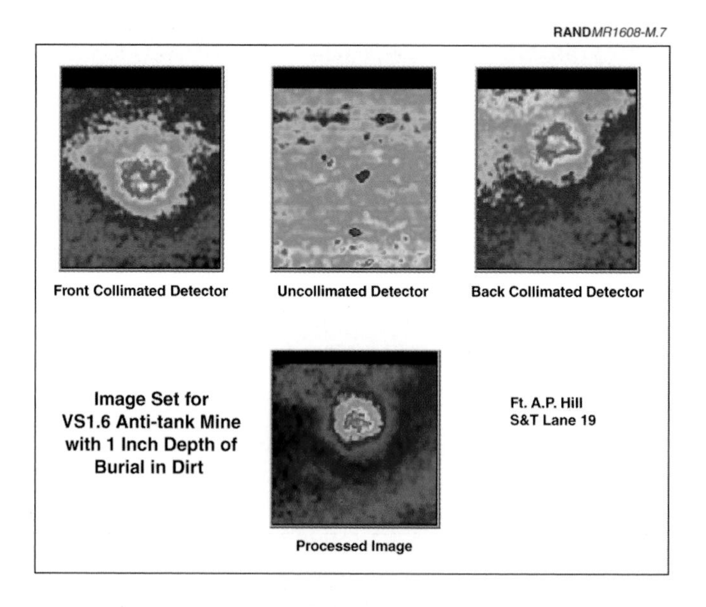


NOTE: The mine is visible in the collimated detector images and, when the surface image in the uncollimated version is subtracted and the mine image is enhanced, the buried mine is the only significant feature in the resulting image.

**Figure M.7—Image Set No. 1 Obtained During the Fort A. P. Hill Test**

## CONCLUSIONS AND EXPECTATIONS

The UF laboratory LMR landmine imaging results shown herein are typical and demonstrate a high degree of selective detection of landmines to several centimeters of burial in homogeneous soil with a plane, featureless surface. Other sets of measurements (results not included here) have shown the effective application of image subtraction to remove substantial soil surface feature mine image cloaking, e.g., Wehlburg et al. [5] and Wehlburg [6]. These references additionally include the laboratory results of mine image degradation due to water and iron additions to the soil. High water content (greater than 25 weight percentage) is required for substantial degradation, but some naturally occurring high iron content soils (greater

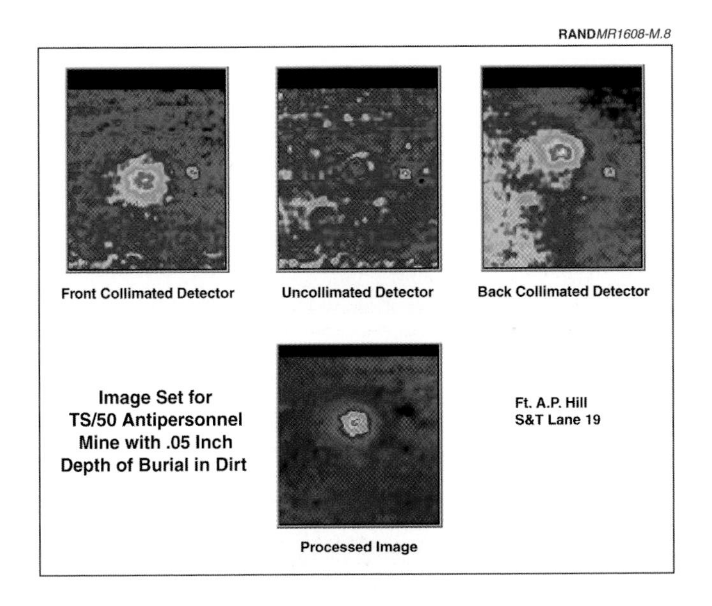


NOTE: The object mine has an essentially flat, featureless surface. However, a bright spot appears in all three detector images and is totally removed in the processed image. This small soil surface feature is a plastic mine-position marker on the soil surface.

**Figure M.8—Image Set No. 2 Obtained During Fort A. P. Hill Test**

than 25 weight percentage) make the LMR technique ineffective. These laboratory measurement efforts have also demonstrated the impact of relatively small design changes for optimal imaging at varying mine depth of burial. The XMIS is optimized for 3-cm mine burial. More information on the XMIS design and the results of the Fort A. P. Hill test is available in Su [7] and Dugan et al. [8], respectively.

The field test LMR landmine detection system (XMIS) has fixed geometric parameters. However, included in the design are some features that were intended to yield information for developing better future designs—e.g., note the two sizes of collimated detector panels employed. Such information, along with extensive Monte Carlo

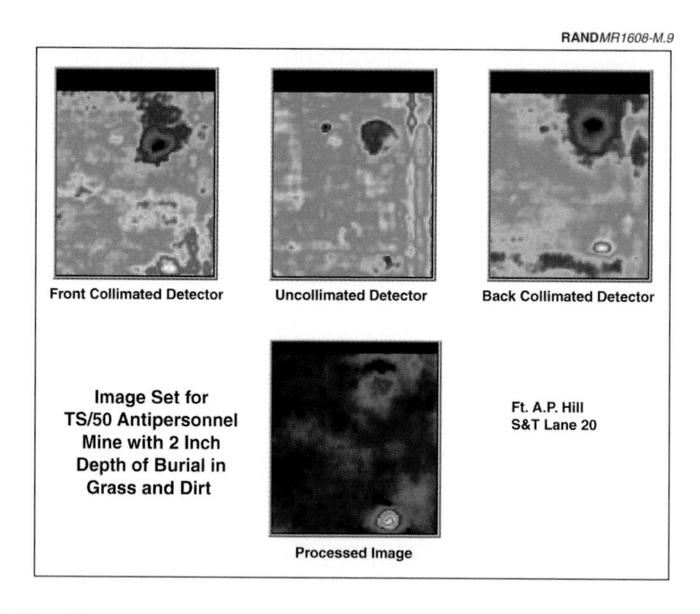


NOTE: The object mine is covered with natural foliage and rocks. Note the large subsurface rock feature near the top of all three detector images. In addition to being of low (rather than high) intensity, this feature is substantially removed along with essentially all soil surface features in the processed image. The mine image is clearly visible at the bottom of the interrogated region (but is not centered due to surface marker confusion for this mine site).

**Figure M.9—Image Set No. 3 Obtained During Fort A. P. Hill Test**

numerical calculation simulations (e.g., Dugan et al. [9]) and some additional conjecture, form the basis for a possible, practical prototype system suggested in the following discussion. Based on the single, very limited test on the mine lanes at Fort A. P. Hill, the mine detection performance parameters of XMIS for depths of burial of 5 cm or less are estimated as: detection probability = 1.00, false alarm probability = 0.03, false alarm rate = 0.10 per square meter. These values are for a soil surface interrogation rate of 1 sq m per minute, and the tests were accomplished in the mode of mine presence confirmation rather than for initial detection because the image sites were specified. The image results, especially in a case like that shown in Figure M.9 (where the site location was only vaguely specified), lend credence to the consideration of the LMR method as a mine presence detection process.

The detection limitation on mine depth of burial (demonstrated in laboratory less than 8 cm, in field test less than 5 cm), the large XMIS weight (150 kg) and size (1.5 m × 1.5 m on scan plane, 1.0 m in height), as well as component ruggedness are major concerns. These weaknesses (in XMIS) are correctable and, to a substantial extent, are the result of limitations in funding and the nature of a device developed in a university setting.

The XMIS fragility, clearly evident in Figure M.10, is easily corrected. It should be noted that light leaks that developed in the relatively frail scintillation/photo-multiplier (PM) detector panels are to a large extent responsible for a reduced performance of the system (and thereby, the order of magnitude increases in electric energy per pixel). In addition, the scintillator thickness in all detectors can be reduced by one-half and their length reduced by one-third without adversely affecting image quality. These changes will yield a total detector weight reduction factor of two-thirds. The large amount of lead shielding, evident in Figures M.10 and M.11, will be substantially reduced by the use of smaller detector panels, but also by employing reduced volume x-ray generators (commercially available now, or designed for this application). The overdesigned universal steel-frame structure of XMIS is easily reduced in weight. These suggested modifications would reduce the total weight of XMIS to about 80 kg and the system dimension in the rotating collimator scan direction to about 1 m.

RANDMR1608-M.10

NOTE: Photograph of XMIS showing most components: x-ray generator with rotating collimator assembly; belt-drive for the rotating collimator; two of the three detector panels (one collimated toward foreground, one uncollimated toward background) with end-mounted photo-multiplier assemblies; lead panels for x-ray shielding; part of the steel channel support structure; x-ray generator controls and high voltage supply (box in background).

**Figure M.10—Components of XMIS**

RANDMR1608-M.11

**Figure M.11—XMIS in Use on the Fort A. P. Hill Mine Lanes**

Each detector panel assembly now has a PM on each of its two ends for light sensing. The location of these PM tubes makes them vulnerable to physical damage and yields inefficient light collection for these scintillator blocks. Replacement with PIN diodes along the entire detector length substantially improves both shortcomings and will yield improved system imaging performance.

The rotating collimator assembly is now moved by a low-cost belt drive with resulting slip and jitter if not in precise adjustment. Such occurrences led to random artifacts in the acquired images that are not corrected. A well-designed gear-driven collimator assembly will solve this problem. In addition, the length of the detector collimators combined with the height of the detector panel plane above the soil surface has a profound effect on the quality (especially contrast) of the collimated detector images. The optimum dimensions depend on the mine depth of burial. In XMIS, the collimator lengths are fixed, and the entire system height is varied by imprecise and cumbersome jacks attached to the trailer frame as shown in Figure M.11. Both collimator lengths and detector plane height above the soil surface can be adjustable and motor-driven. Accumulated image information during a soil region scan with nominal position settings can be employed to reset dimensions for optimal mine imaging once presence of a mine is suspected.

It is expected that applying kVp-modulation to the x-ray generator, such as reported by Towe and Jacobs [2], will substantially enhance the discernibility of a landmine image to the extent that field application will yield mine detection to depth of burial of 10 cm or more. This feature addition is crucial to the solution of the limited depth of burial mine detection sensitivity, but it is well within current technology. If the rotating collimator-induced scan is maintained in future designs, smaller x-ray generator heads are available and should be employed. This will lead to further system weight and height reductions. All large mechanical motions of the XMIS scanning mechanisms can be eliminated and system ruggedness vastly improved if a concept (already conceived and tested by Bio-Imaging Research Inc.) is included in a future design. The new x-ray generator concept includes a multicathode linear array along the axis of a single liquid-cooled anode tube. The cathodes are fired sequentially to achieve one direction of scan. Coordination with a segmentation of the detector panel activation could increase, by an order of magnitude, the scan speed by simultaneously illuminating various portions of the object. The other scan direction can be achieved by a small angle mechanical rotation of the generator tube near the anode axis. It should be mentioned that the availability of such an x-ray generator should yield relatively easy to achieve extension of scan dimensions (along the anode tube direction) of up to 3 m, such that the military application to tank lane mine detection becomes plausible.

The system modifications discussed above, with the exception of the multicathode generator development, could be completed in one year at a total cost of $500,000. The multicathode generator development could be attempted in about two years at a total cost of $1.5 million. Substantial UF–industry collaboration is assumed in these estimates. It is difficult to estimate the cost of a system once this development process has been completed. If conventional, single focal spot x-ray generators are employed, total system cost should be less than $50,000, and with significant manufactured quantity, less than $30,000.

## REFERENCES

1. Odeblad, E., and A. Norhagen, "Electron Density in a Localized Volume by Compton Scattering," *Acta Radiologica*, No. 45, 1956, pp. 161–167.

2. Towe, B., and A. Jacobs, "X-Ray Backscatter Imaging," *IEEE Transactions on Biomedical Engineering*, BME-28, 1981, pp. 646–654.

3. Roder, F., and R. Van Konyenburg, *Theory and Application of X-Ray and Gamma-Ray Backscatter to Landmine Detection*, Fort Belvoir, Va.: U.S. Army Mobility Equipment Research and Development Center, Report 2134, 1975.

4. Su, Z., A. Jacobs, E. Dugan, J. Howley, and J. Jacobs, "Lateral Migration Radiography Application to Landmine Detection Confirmation and Classification," *Optical Engineering*, Vol. 39, No. 9, 2000, pp. 2472–2479.

5. Wehlburg, J., S. Keshavmurthy, E. Dugan, and A. Jacobs, "Experimental Measurement of Noise Removal Techniques for Compton Backscatter Imaging Systems as Applied to the Detection of

Landmines," in *Detection Technologies for Mines and Minelike Targets,* A. C. Dubey, R. L. Barnard, C. J. Lowe, and J. E. McFee, eds., Seattle: International Society for Optical Engineers, April 1996.

6. Wehlburg, J., *Development of a Lateral Migration Radiography Image Generation and Object Recognition System,* Ph.D. dissertation, University of Florida, 1997.

7. Su, Z., *Fundamental Analysis and Algorithms for Development of a Mobile Fast-Scan Lateral Migration Radiography System,* Ph.D. dissertation, University of Florida, 2001.

8. Dugan, E., A. Jacobs, Z. Su, L. Houssay, D. Ekdahl, and S. Brygoo, "Development and Field Testing of a Mobile Backscatter X-Ray Lateral Migration Radiography Landmine Detection System," in *Detection and Remediation Technologies for Mine and Minelike Targets VII,* J. T. Broach, R. S. Harmon, and G. J. Dobeck, eds., Seattle: International Society for Optical Engineering, April 2002.

9. Dugan, E., S. Kesharmurthy, A. Jacobs, and J. Wehlburg, "Monte Carlo Simulation of Lateral Migration Backscatter Radiography Measurements," *American Nuclear Society Transactions,* No. 76, June 1997, pp. 140–142.

Appendix N

# NEUTRON TECHNOLOGIES (PAPER I)

*John E. McFee, Canadian Centre for Mine Action Technologies*[1]

## INTRODUCTION

Detection of landmines using nuclear techniques has been studied extensively since the late 1940s. Nuclear techniques look at either a return radiation, which is characteristic of explosive components that are infrequently found in soil (e.g., nitrogen or carbon), or an intensity change of a noncharacteristic scattered radiation, which is a function of a parameter that differs between soil and explosives. Noncharacteristic radiation methods are essentially anomaly detectors; that is, they detect inhomogeneities in the medium and inclusions in addition to mines. Virtually every conceivable nuclear reaction has been examined, but after considering a number of factors (many of them linked)—including selectivity, sensitivity, probability of detection, false alarm rate, soil absorption, time to make a detection, limitations due to fundamental physics, and technical limitations (size, weight, power, present and future availability of sources and detectors)—only a few have potential for mine detection. Probably the most thorough examination of nuclear reactions for landmine detection is the report by Coleman et al. [1] sponsored by the U.S. Army Mobility Equipment Research and Development Center (now called the Night Vision and Electronic Sensors Directorate). A workshop was held in 1985 to revisit the conclusions of the Coleman report in light of advances in technology and to identify any nuclear

[1]This paper is based on research conducted at the Canadian Centre for Mine Action Technologies, under the sponsorship of the Canadian government (Defence R&D–Suffield).

techniques that should be developed for mine detection [2]. Experts concluded that the most promising nuclear technologies were, in order, x-ray backscatter imaging, thermal neutron capture gamma rays, neutron thermalization (moderation), and differential collimated photon scattering (now called x-ray lateral migration). Coleman did not do an explicit ranking of reactions; however, the top three techniques are the same for both studies. Moler did a further reassessment in 1991 [3] in light of research that had been done as a result of his 1985 report. His conclusions remained the same. Other reviews have included nuclear methods in the context of unexploded ordnance and landmines, [4] military demining [5], and humanitarian demining [6], all with conclusions similar to those of Coleman and Moler.

The following discussion is restricted to reactions that involve excitation by, or emission of, neutrons. For a neutron reaction to be potentially adaptable to handheld antipersonnel mine detection, it must have a large cross section and be amenable to using very efficient detectors, so that relatively weak sources can be used. This allows a decrease in size and weight of shielding for electronics and personnel. The only feasible reaction in this category is neutron moderation. We shall nevertheless look at the other neutron reactions. Although humanitarian demining is mainly concerned with the detection of antipersonnel mines, in some scenarios handheld use is not required and in others, such as supply route proving, vehicle-mounted detection of antitank mines may be needed.

## NEUTRON REACTIONS

Because of range limitations in the soil and mine, the number of neutron reactions can be dramatically reduced by eliminating those that involve charged particle excitation or detection. The remaining reaction classes are neutron excitation/photon detection (neutron capture gamma rays, neutron inelastic scattering gamma rays, neutron activation); neutron excitation/neutron detection (neutron moderation, neutron elastic resonance scattering, neutron inelastic scattering); and photon excitation/neutron detection (photoneutron emission).

## Neutron Capture Gamma Rays (Thermal Neutron Analysis)

Many of the materials in soils and landmines emit gamma rays when thermal neutrons are captured. Most research since the early 1950s has concentrated on the 10.835 MeV transition in nitrogen because bulk nitrogen is indicative of the presence of explosives,[2] there are virtually no nearby competing reactions except the weak 10.611 MeV transition from $^{29}Si$, and it is sufficiently isolated so that low energy resolution NaI(Tl) detectors may be used. (High-resolution intrinsic Ge detectors are an alternative, but they are much more expensive for similar efficiency and less suitable for fieldable systems because they are less rugged, microphonic, more prone to neutron damage, and must be cryogenically or electromechanically cooled.) It has been estimated from experiments and modeling [7] that neutron capture detectors could be used in a scanning role (95-percent probability of detection [PD] with false alarm rate [FAR] of less than 1 per 500 meters at speeds of at least 10 km per hour). This is contradicted by other independent analyses [3,8].

Researchers have used a variety of isotopic and electronic sources and detectors. For confirmation detection, high-intensity sources (greater than $10^8$ neutrons/second) are required. Performance results from low-intensity source systems, where silicon interference limits performance, cannot easily be extrapolated to those with high-intensity sources, where pulse pileup is the limiting factor.

Defence R&D Canada (DRDC) and Bubble Technology Industries (BTI) have been developing a thermal neutron analysis (TNA) detector as a confirmation sensor on a teleoperated, multisensor vehicle-mounted mine detector (Improved Landmine Detector Project [ILDP]) since 1994 [9]. To be usable for confirmation mine detection, a TNA detector must be carefully designed with its role specifically in mind. This was done, paying particular attention to electronic and shielding design. The production version is based on an intense $^{252}Cf$ source and four NaI(Tl) detectors. The sensor head is roughly a cube with 0.6-m sides and has a mass of about 216 kg. The electronics consume about 200 watts. A prototype 14.8 MeV DT [deuterium/tritium] neutron generator-based version, with a roughly 10-percent-

---

[2]Explosives in landmines contain about 18–38 percent nitrogen by weight; soil contains less than 0.07 percent.

larger sensor head, consuming 1 kW, has been built and is currently undergoing testing [10]. Preliminary tests supported by calculations suggest that it should perform at least as well as the isotopic TNA—and possibly better. The lab (nonproduction) isotopic source prototype TNA has functioned well as a part of ILDP and in standalone tests in stringent climatic conditions, including in temperatures from –20°C to +40°C, snow, rain, dust, and extreme dryness.

Only the nonproduction prototype isotopic source DRDC TNA has had its performance quantified. The production version is currently undergoing testing and is expected to perform at least as well. In nonblind testing at DRDC-Suffield, it was found that the TNA can confirm the presence of a variety of surface-laid or shallowly buried antitank mines in a few seconds to a minute, depending on the explosive mass [9]. It is also capable of confirming the presence of antitank mines buried to a depth of 20 cm and shallow, large (greater than 100 g nitrogen) antipersonnel mines in less than five minutes. Limited blind testing conducted by the Institute for Defense Analyses as part of the 1998 Ground Standoff Mine Detection System trials showed that the prototype was capable of a PD/probability of false alarm (PFA) of 0.79/0.00 at Aberdeen Proving Ground, Md., and 1.00/0.32 at Socorro, N.M. [11]. The TNA is capable of better performance; however, the laboratory prototype tested was not intended to be a rugged, fielded system, and environmental conditions were unfavorable at both sites (temperatures from +35°C to +40°C). More significantly, software and hardware shortcomings, which have since been rectified in the production version, decreased performance.

The Canadian TNA program appears to be the only currently active one. Present research and development (R&D) involves detector and signal-processing improvements to decrease count times and increase sensitivity. These are expected to be incremental improvements. Extensive system characterization, planned for this year, is essential for both the isotopic- and neutron generator–based detectors. A preliminary study, started in September 2002, is investigating the feasibility and benefits of adapting the Canadian TNA detector system to combine TNA with fast neutron analysis (TNA/FNA).

Numerous experimental and modeling studies and reviews since the 1950s and experience with the Canadian TNA all point to the fact that TNA is feasible only for vehicle-mounted confirmation detection of antitank and large antipersonnel mines. In spite of occasional claims, a practical, lightweight, and person-portable system is unachievable, based on the physics, as opposed to the technology. No order of magnitude breakthroughs are expected.

## Neutron Inelastic Scattering Gamma Rays (Fast Neutron Analysis)

In this method, fast neutrons excite soil and mine nuclei by inelastic scattering. As the nuclei de-excite, they emit characteristic gamma rays. The only practical source at present is a 14.8 MeV DT neutron generator. At that energy, nitrogen cross sections are much smaller than for thermal neutron capture gammas and offer no clear advantage over the latter. Production of 4.44 MeV photons from $^{12}C(n,n'\gamma)$ has a large cross section, but organic and carbonaceous material will cause false alarms.[3] Efficient, robust detectors, such as NaI(Tl), lack sufficient energy resolution to distinguish 4.44 MeV photons from Si and O background gammas, and, as stated above, Ge detectors are not practical.

The main claimed advantage for the neutron inelastic scattering gamma ray method is that it could be used to discriminate explosives from organic soil materials by measuring the C:H:O concentration ratios in the interrogated volume [12]. Given the cross sections, it seems that the time required to obtain sufficiently accurate ratios for reliable discrimination would be much longer than TNA interrogation times. Further, repeated studies since the late 1950s have consistently concluded that landmine detection through neutron inelastic scattering gamma production is not at present practical with available sources and detectors. Given these elements, it is difficult to see any advantage over thermal neutron capture.

It has been implied that a neutron inelastic scattering gamma ray detector for landmines can be easily constructed from components used in other roles, such as coal slurry analysis [12]. However, quite the opposite is true. Experience with TNA for mine detection shows that detector systems must be purpose-built for the role with intense

---

[3]Explosives in landmines contain about 16–37 percent carbon by weight; soils contain about 0.1–9.0 percent.

neutron sources and carefully designed shielding. The intensity of sources required precludes the use of the associated particle technique to acquire position information. There may be some advantages to combining TNA with FNA, such as adding a carbon detection capability to the nitrogen capability of TNA. However, this must be done in a purpose-built design. DRDC has recently initiated a feasibility study to examine if the DRDC/BTI neutron generator source TNA can be modified to provide an FNA capability.

Changing the neutron energy to 6 MeV has the advantage of producing a 5.10-MeV nitrogen gamma with no interfering gammas from soil. Currently, there are no sources of 6-MeV neutrons, and a feasibility study to determine if one could be constructed is recommended.

## Neutron Activation

Fast neutrons can activate materials in soils and mines through a number of reactions. These materials can be detected by measuring characteristic gamma rays from their radioactive daughters. Half-lives of these radioactive components vary from milliseconds to days. Silicon and oxygen account for the dominant gamma rays from soil, but detecting a lack of these materials is equivalent to detecting voids. The $^{14}N(n,2n)$ reaction will produce characteristic 511 keV gamma rays, but they are much weaker in intensity than the gammas from silicon. A series of experiments from the 1950s through the 1970s verified that the nitrogen signal is far too weak to be usable for practical scan/dwell times. Silicon voids were detectable but only with very intense neutron generators (about $2 \times 10^{10}$) neutrons/second, five minute irradiation) and at very slow speeds. These results are limited by the physics and preclude the development of a practical detection system.

## Neutron Moderation

This method involves irradiating an area with fast neutrons and detecting subsequently moderated and returned slow neutrons. Explosives contain 2–3 percent hydrogen by weight, while soils may contain 0-percent to more than 50-percent hydrogen. Thus the presence of an anomaly in the measurement of hydrogen density may be used to imply the presence of a mine. Measurement of the albedo signal (ratio of number of slow neutrons returned from the soil to the number of incident fast neutrons) is then used as an indicator of the presence of a mine. The large cross sections make it amenable in principle to handheld operation and detection of antipersonnel mines. The chief limiting factor in the signal-to-clutter ratio (SCR) is hydrogen in groundwater. The hydrogen densities of the soil and the mine are equal, and mines cannot be detected when the gravimetric percentage of water is between 18 and 27. Other factors that affect the SCR are ground surface irregularities and detector height variations. In practice in the past, these combined factors have rendered this detection technique, using nonimaging detectors, useless in all but the driest of conditions. One method of reducing false alarms from all of the above sources is to spatially image the neutrons coming from the ground.

The U.S. Army sponsored research in neutron moderation detection of landmines from the early 1950s through the early 1990s. Isotopic sources, such as Po-Be and Cf, with typical outputs of $10^6$ neutrons/second, were used as well as accelerator sources employing different reactions yielding 1.1 MeV, 2.8 MeV, and 14.8 MeV neutron energies. Detectors have included $BF_3$ and $^3He$ proportional counters and $^6LiI$ crystals wrapped in Cd. All previous research has been nonimaging. A few research groups are currently involved in neutron moderation studies involving simulation and experiment. Researchers active in this area include Frank Brooks's group at the University of Capetown, South Africa [13]; Carel W.E. van Eijk's group at the Delft University of Technology in the Netherlands [14]; and Julius Csikai's group at the Institute of Experimental Physics, University of Debrecen, Hungary. For neutron moderation to be useful, imaging must be employed to reduce the high FAR. None of the previously mentioned work involves developing imaging systems.

A DRDC/BTI team has been developing a neutron moderation imager based on a large scintillation screen and wavelength shifter optical fiber readouts for mine detection since 2000 [15]. A proof-of-concept detector is nearly constructed, and preliminary testing will be completed by the project's end in March 2003. The DRDC/BTI imager will be 50 cm × 50 cm, with a mass of roughly 13 kg and a power consumption of 10 watts. Various other imaging technologies can be investigated, but these, together with clever packaging, will likely lead to only incremental decreases in mass and power. There is little published data on measured PD and FAR, which are largely a function of how long one is willing to count. It is unlikely that performance will dramatically improve over the estimates from simulation studies [14]. These showed that images of suitable fidelity could be made of antipersonnel and antitank mines. Detection of a small antipersonnel mine (PMA2) with a PD/PFA of 0.98/0.02 could be achieved in about 6 seconds at a 5-cm depth in sand with 10-percent water. At a depth of 10 cm, the time increases to 21 seconds. For a small antitank mine (500-g explosive, 7-cm diameter, 7-cm height [cylinder]) at a 5-cm depth in sand with 10-percent water, the detection time decreases to 0.4 seconds. It appears that detection in environments with water contents between 10 percent and 20 percent may be possible.

Neutron moderation imaging may have potential for quick confirmation or slow scanning of antipersonnel mines in soils with moisture content less than or equal to 10 percent and possibly between 10 percent and 20 percent. A prototype Canadian instrument will be ready by March 2003, and its performance will be extensively characterized in summer 2003. No other countries appear to be actively investigating neutron moderation imaging at this time, although the Delft group recently reported having a design for an imaging detector based on position-sensitive $^{3}$He tubes. Although the Canadian instrument could use a distributed array of low-intensity, point $^{252}$Cf sources, better performance will be achieved with a uniform continuous source. Although none is available at present, such a source is being developed by DRDC and will be installed in the detector by March 2003. A further improvement might be gained by employing a distributed pulsed neutron source. The thermal decay constant for explosives is significantly different than that for either wet or dry soil. Measurement of the decay constant for each pixel could significantly improve the contrast between mine and soil, particularly in wet soil. Richard Craig of Pacific Northwest National Laboratory is developing a nonimaging detector based on this method. Distributed, wide area pulsed neutron sources do not exist, but DRDC and BTI have some design concepts, and a funding proposal to develop one has been submitted.

## Neutron Elastic Resonance Scattering

In light elements, such as those in explosives and soil, the cross sections for elastic scattering of neutrons in the keV to MeV range exhibit resonances as a function of neutron energy. Each element has a characteristic set of resonance energies and intensities. Because the resonances are generally narrow and sit on a broad continuum component of the cross section, it is necessary to use an accelerator, such as a Van de Graaff, to produce monoenergetic neutrons. This is not practical in the field. Broad-spectrum sources (e.g., fission sources) are practical, but the resonance component of the cross section is generally much smaller than the continuum component, which is common to both explosive and soil materials. Thus, no measurable difference is observed between the target and background. Experiments conducted in the early 1950s using a carbon target, a continuous source, and a carbon filter confirmed this. High-energy-resolution detectors, such as time-of-flight (TOF) detectors or $^3$He ionization chambers, are unlikely to assist in resolving the continuum component from the resonance component. TOF detectors are too large and inefficient. Recoil broadening is much greater than the resolution of either detector type and hence averages over the narrow resonances. The detector resolutions are also greater than or equal to the resonance widths. This leads to the inability to resolve resonances from the continuum to which all elements contribute. Thus, several close resonances in O and N cannot be distinguished from each other or the continuum, and the two narrow carbon resonances cannot be distinguished from the continuum. This makes the method nonspecific, and there appear to be no productive avenues for further research in this area.

## Neutron Inelastic Scattering

When neutrons scatter from certain nuclei, they can lose energy by exciting the nucleus from its ground state to an excited state. In principle, the nucleus can be detected by detecting an excess of neutrons with an energy that is lower than the incident energy by the difference between the ground and excited states. Carbon's large cross section to the 4.4-MeV level makes it the most likely candidate, although a carbon detector is prone to false alarms from organic materials and carbonaceous soil. Nitrogen, which would be more specific to explosives, has cross sections that are about three to eight times smaller. The only practical monoenergetic source is a 14.8-MeV neutron generator. In principle, scattered continuous fission spectra could be unfolded to reveal the peaks in the inelastic scattering neutron spectrum, but the continuum of neutrons from direct and multiple scattering, including inelastic scattering from low-lying states, in background material nuclei would make this method impractical. Neutrons are around 10 MeV for carbon and 7–13 MeV for nitrogen. The only practical detectors in this range are TOF systems, which are inefficient and large. For scanning, neutron fluxes of $10^{11}$ neutrons/second would be required, while for confirmation, $10^9$ neutrons/second would likely be adequate. At present, there are no advantages—and some disadvantages—to this technique over neutron inelastic scattering gamma rays. Any system would be too large and heavy for humanitarian demining.

The state of progress in the development of high-efficiency neutron spectrometers in the 7–13 MeV range should be examined to determine if development is feasible and practical.

## Photoneutrons

If incident gamma-ray energies exceed the Q-value for neutron production, neutrons will be emitted with energies approximately equal to the gamma-ray energy minus Q-value minus the energy of the excited state in which the resultant nucleus is left (a correction for the nuclear recoil is required). Detection of neutrons without energy discrimination has been attempted using a bremsstrahlung beam whose endpoint energy was slightly higher than the 10.5 MeV $^{14}N$ threshold. It failed because of the strong background of neutrons from silicon. Detection of $^{13}C$ and $^{2}H$ is also possible; if an incident energy of 6 MeV were used, only those two isotopes would be excited. However, this would be prone to false alarms from organic materials and carbonaceous soils, and down-scattered fast neutrons from water would dominate the response.

If a suitable energy-sensitive neutron detector were used, the spectrum of emitted characteristic neutrons might be used to identify the nucleus. The normal detectors in the 100 keV to few MeV range are proton recoil and TOF. The former generally has poor energy resolution due to uncertainties in unfolding its flat response to monoenergetic neutrons. The latter is very large if reasonable energy resolution is desired. A high-resolution, fast neutron spectrometer based on a $^{3}$He ionization chamber has been used for photoneutron spectroscopy in this energy range [16] and was proposed in this role for mine detection [4]. These instruments are no longer manufactured, but DRDC has such a detector. Plans are under way to repeat the previous bremsstrahlung threshold experiments over mines and soil at various endpoint energies from 6 to 14 MeV in the near future. A Linac will be used as a source and a $^{3}$He ionization chamber, proton recoil counters, and threshold energy bubble detectors will all be used for comparison.

The required gamma-ray energies are too high for isotopic sources and a bremsstrahlung source, such as a medical Linac, is the only practical alternative. Thus, a detector based on photoneutron spectroscopy would be too big for humanitarian demining. It might be feasible for a limited role as a confirming detector for route proving, but it is not likely to offer any advantages over neutron capture gamma rays.

## SUMMARY AND R&D RECOMMENDATIONS

Among neutron reactions, only neutron capture gamma rays have so far led to fielded systems. However, because of the physics, these detectors are vehicle-mounted and are restricted to detecting antitank and large antipersonnel mines; they are not practical for detection of small antipersonnel mines. The only neutron technology that may be feasible as a handheld detector of small antipersonnel mines is neutron moderation imaging. Such detectors would function in a confirmation or slow scanning mode and would not work in very wet soil. R&D on TNA and neutron moderation imaging is presently concentrated in Canada.

No neutron-based technologies will provide an order of magnitude improvement in detection speed or efficiency. Modest research programs in selected aspects of neutron moderation imaging, neutron capture gamma rays (TNA), neutron inelastic scattering gamma rays (FNA), and TNA/FNA combinations may be warranted for mine detection. These should complement, rather than compete with, existing R&D programs.

Specific suggestions follow. Estimated person years (PYs) and funding follow each project item in parentheses. These should be treated as rough order of magnitude estimates only.

Desirable R&D for humanitarian demining (detection of anti-tank/large antipersonnel mines for route proving):

- A feasibility study to determine if a practical 6-MeV neutron generator could be constructed for FNA (1 PY, $200,000).
- Design and construction of a practical 6-MeV neutron generator (subject to positive outcome from feasibility study) (6 PYs, $5 million).
- Feasibility study to determine whether high efficiency and resolution neutron spectrometers in the 7–13 MeV range could be developed (1 PY, $200,000).
- Design and construction of high efficiency and resolution 7–13 MeV neutron spectrometers (subject to positive outcome from feasibility study) (5 PYs, $200,000).

## REFERENCES


1. W. A. Coleman, R. O. Ginaven, and G. M. Reynolds, *Nuclear Methods of Mine Detection Vol. III*, Science Applications Inc., Technical Report SAI-74-203-L, May 1974.
2. R. B. Moler, *Workshop Report: Nuclear Techniques for Mine Detection Research, Lake Luzerne, NY, July 22–25, 1985*, Army Belvoir Research and Development Center, Technical Report AD-A167968, November 1986.
3. R. B. Moler, *Nuclear and Atomic Methods of Mine Detection (U)*, Catharpin, Va.: Systems Support Inc., AD-A243332, November 1991.
4. J. E. McFee and Y. Das, *Review of Unexploded Ordnance Detection Methods (U)*, Defence Research Establishment–Suffield, Report SR 292, September 1981.

5. J. E. McFee, Y. Das, A. Carruthers, S. Murray, P. Gallagher, and G. Briosi, *CRAD Countermine R&D Study: Final Report (U)*, Defence Research Establishment–Suffield, Report SSP 174, April 1994.

6. J. E. McFee and Y. Das, *A Research and Development Plan for Land Mine Detection Technologies for the Canadian Center for Mine Action Technologies (CCMAT) (U)*, Defence Research Establishment–Suffield, Report TR 1999-022, February 1999.

7. *Feasibility Study of Neutron Activation Techniques for Mine Detection: Final Report (U)*, Santa Clara, Calif.: Science Applications International Corporation, AD-B135540, August 1989.

8. T. J. Jamieson and F. Lemay, *Feasibility and Costing Study for the Use of Thermal Neutron Activation Analysis (TNA) as a Primary Non-Metallic Mine Detection System (U)*, SAIC Canada, DRES Contract Report CR 1999-115, October 1995.

9. T. Cousins, T. A. Jones, J. R. Brisson, J. E. McFee, T. J. Jamieson, E. J. Waller, F. J. LeMay, H. Ing, C. E. Clifford, and B. Selkirk, "The Development of a Thermal Neutron Activation (TNA) System as a Confirmatory Nonmetallic Land Mine Detector," *Journal of Radioanalytical and Nuclear Chemistry*, No. 235, September 1998, pp. 53–58.

10. D. S. Haslip, T. Cousins, H. R. Andrews, J. Chen, E. T. H. Clifford, H. Ing, and J. McFee, "DT Neutron Generator as a Source for a Thermal Neutron Activation System for Confirmatory Land Mine Detection," in *Hard X-Ray and Gamma-Ray Detector Physics III*, R. B. James, ed., Seattle: International Society for Optical Engineering, July 2001, pp. 232–242.

11. J. D. Silk, L. Porter, and R. Moler, *Vehicular Mounted Mine Detector (VMMD) Test of Neutron Activation Technology*, Alexandria, Va.: Institute for Defense Analyses, D-2286, March 1999.

12. P. C. Womble, G. Vourvopoulos, and J. Paschal, "PELAN 2001: Current Status of the PELAN Explosives Detection System," in *Hard X-Ray and Gamma-Ray Detector Physics III*, R. B. James, ed., Seattle: International Society for Optical Engineering, July 2001, pp. 226–231.

13. F. D. Brooks, A. Buffler, and M. S. Allie, "Landmine Detection by Neutron Backscattering," in *7th International Conference on Applications of Nuclear Techniques*, G. Vourvopoulos, ed., University of Western Kentucky, 2001.

14. C. P. Datema et al., "Landmine Detection with the Neutron Backscatter Method," *IEEE Trans. Nucl. Sci.*, Vol. 48, No. 4, 2001, pp. 1087–1091.

15. J. McFee, H. R. Andrews, H. Ing, T. Cousins, A. Faust, and D. Haslip, "A Neutron Albedo Imager for Land Mine Detection," in *Detection and Remediation Technologies for Mines and Minelike Targets VII*, J. T. Broach, R. S. Harmon, and G. J. Dobeck, eds., Seattle: International Society for Optical Engineering, 2002.

16. J. McFee, *Photoneutron Spectroscopy of the Nuclei Near Mass 200*, Ph.D. thesis, McMaster University, Hamilton, Ontario (Canada), 1977.

Appendix O

# NEUTRON TECHNOLOGIES (PAPER II)

*David A. Sparrow, Institute for Defense Analyses*[1]

## EXECUTIVE SUMMARY

Neutrons pass relatively easily through matter and have been used successfully to interrogate the earth for bulk presence of elements, most notably those indicating petroleum. Applying these technologies to explosive detection is challenging because the "bulk" one is looking for can be as small as 100 g or less. The signals from 100 g of material are dwarfed by the background from the rest of the nearby earth and even by cosmic rays. To date, performance at confirming the presence of small amounts of ordnance is poor, even with dwells as long as 10 minutes over areas smaller than a half square meter. This is in part because of variations in the background signal over short distances on the earth's surface, which can either mask or mimic the ordnance signal.

There is no prospect that current approaches can deliver performance sufficient to be of use in humanitarian demining. There is some prospect that an approach based on high-resolution Ge detectors could resolve the gamma-ray signals sufficiently well enough to allow discrimination of the explosive and background signals. To pursue this requires an instrumentation-oriented research program, looking initially at a combination of phenomenology and engineering development. Using cryogenic Ge detectors, possibly augmented with Compton suppression techniques and cosmic-ray vetoing, a

[1]This paper has been reprinted here with permission of copyright owner, Institute for Defense Analyses (IDA).

high-resolution spectrum could probably be produced that would allow for discrimination into constituent elements. A fairly small number of institutions could pull this off.

We regard such a research program as a high-risk undertaking for humanitarian demining application. A system could probably be produced that would yield a spectrum highly enough resolved to allow for discrimination. However, to develop a version of such a system that would even match current performance, much less speed it up by a factor of 10, would require an enormously expensive system, which would probably require great technical sophistication to operate. A disciplined investigation in this area is more likely to yield either systems or insights of value to the unexploded ordnance (UXO) community, where much larger ordnance items are often of interest.

## GENERAL DISCUSSION

Neutrons pass relatively easily through matter but do interact with atomic nuclei. These interactions produce a set of resulting gamma rays (and sometimes of charged particles as well) that can provide a distinctive signature of the nucleus and hence of the chemical element with which they are interacting. Applications of this interrogation technique have been made with both thermal neutrons from radioactive sources and energetic and thermal neutrons from neutron generators.

Because neutrons have a long mean free path in matter, these techniques are best suited to bulk interrogation, "bulk" meaning on the scale of a ton. The most common and most successful application is in oil exploration, where underground neutron generators and gamma detectors have been used for years to confirm the presence of significant oil deposits in the earth's crust. The primary signature is the presence of relatively large amounts of carbon.

The analogous signature for explosive detection is relatively high amounts of nitrogen. However, to be useful in humanitarian demining, any of the chemical-specific techniques must exhibit high performance against essentially any explosive in quantities of 100 g. In other words, instead of looking for changes in composition of the soil of several percent, one is looking for changes at the level of 100 ppm.

The small quantities involved lead directly to the challenges for this technology in humanitarian demining. First, the signal received is overwhelmingly from the earth's crust, secondarily from cosmic rays, and only slightly from the target of interest. Thus, the signal of interest must be distinguished from counting fluctuations in the earth and cosmic-ray background, and true spatial variation in the signal from the earth, and temporal variation in the signal from the sky. In other words, the interference comes in two types. First, from fluctuations, which can be noise-like and therefore are, in principle, reducible relative to the signal of interest by counting for long times. (See quantitative discussion #1 in the next section.) Second are variations in the earth's crust that lead to signals that can either mimic or mask the signal of the ordnance. These are a real effect on the ground and cannot be eliminated by counting for longer times. With the relatively low-resolution detectors currently in use, there is no prospect that adequate performance can be achieved in confirming or refuting the presence of 100 g of explosive, even given very long counting times. Using high-resolution detectors might allow discrimination at the chemical element level, leading to adequate performance. High resolution would also reduce backgrounds, thereby shortening counting times. (See quantitative discussion #2 in the next section.)

Two types of sources are proposed for this sort of work—radioactive fission or alpha particle sources, which produce low-energy neutrons that are quickly thermalized, and neutron generators that produce high-energy neutrons. If the latter is used, the possibility exists to look for signatures with the beam on from (n,n') and (n,p) reactions, with the beam off from thermal neutron capture (n,gamma), and finally from neutron activation of "long-lived" (greater than 1 second) nuclear states. All these processes suffer from the problem of masking and mimicking discussed above. In general, the situation is much more complex and the backgrounds harder to understand as the energies of the gammas get lower than in the 10.8-MeV case considered below.

The only prospect we envision for separating the targets of interest from small variations in the background would involve a very expensive upgrade in detector technology, from scintillator materials to solid-state devices. This would permit separation of the gamma-ray peaks of different elements, and consequently the background composition could in principle be determined for each location, rather than assumed to be equal to that of some nearby location and subtracted.

The downside of using solid-state detectors is cost and complexity. The detectors require cryogenic cooling. In addition, the needed background suppression may require Compton suppression and/or cosmic ray vetoing.

Developing a system that would do all this, and taking data and analyzing it to determine how good the resolution required for the spectrum must be to yield directly the elemental abundances, would be at least a Ph.D.-level thesis undertaking, with possibly several such undertakings. A limited number of universities could support such research. This would still not yield a system but could yield a test bed to provide the information necessary to assess optimal performance and to support design of a system. Scientific or instrumentation questions include: How good must the resolution be to discriminate elements? Is Compton suppression necessary? Is cosmic-ray vetoing necessary? If background variation is controlled, how long would one need to count for a given signal-to-noise ratio (SNR)? Engineering questions include: What sort of source is needed? Are all three processes—fast neutron, thermal neutron and activation—needed? How is the shielding, and radiation safety generally, to be accomplished? This undertaking would last three to five years and would cost at least $5 million.

Although such a research project is supportable, we believe it to be high risk in the context of achieving dramatic improvement in the speed of humanitarian demining. Even if the sensor and processing can support discrimination of the spectrum in principle, in practice there will remain the necessity to count for long periods of time, unless an enormous investment is made in buying high-resolution detectors. In addition, the system would be of great mechanical, electrical, and computational complexity if it used a neutron generator. If it used a radioactive source, additional complexity would result from the need for careful handling of radioactive materials near explosives. It is difficult to imagine the path required for successful technology transition in the humanitarian demining application. Nevertheless, such a research program might prove of use to the

UXO community, where there is often interest in confirming the presence of much larger amounts of explosive.

## QUANTITATIVE EXAMPLES

We present some sample calculations based on thermal neutrons from a $Cf^{252}$ source developed in the mid-1990s. Results would vary for a differently engineered system or for a concept that looked for lower-energy gammas. These examples are nonetheless illustrative of the challenges.

### 1. Improving Performance of Current Technology by Counting Longer

A simple example involves the use of thermal neutrons to detect $N^{14}$ via the 10.8-MeV capture gamma ray. The primary background is from $Si^{29}$ in the soil, on which neutrons capture to a 10.6-MeV gamma ray. For a particular NaI detector, the total background is about 100 counts per minute, 60 of which are due to Si in the soil, 20 to cosmic rays, and 10 each to energetic neutrons from the Cf fission and pileup. (The pileup is significant because it means that improvements from simply increasing source intensity will be minimal.) One hundred grams of explosive would add about three counts per minute. Considering only counting noise, we have, for time in minutes:

$S = 3t,$

$N = (100t)^{1/2},$

which implies that a 12-detector system would have a signal to noise of approximately $SNR = (1.08t)^{1/2}$, and need more than 20 minutes to achieve an SNR of 5.

Unfortunately, even an SNR of 5 does not guarantee good performance because small fluctuations in the amount of $Si^{29}$, or the depth at which high silicon layers were found, would easily mask (if the silicon was less than expected or deeper) or mimic (the silicon was more than expected or shallower) the explosive.

## 2. Improving Performance with Higher-Resolution Detectors

Using detectors that resolved the N and Si peaks would eliminate the problem of Si fluctuation and might further reduce the overall background by an order of magnitude. This suggests that a little more than two minutes counting time would be needed to achieve an SNR of 5, assuming the overall system efficiency could be preserved with the smaller, more expensive Ge detectors. The combination of background reduction for shorter counting time and improved discrimination for acceptable performance are necessary—but may not be sufficient—for a useful system.

## BIBLIOGRAPHY

Grodzins, L., "Nuclear Technologies for Finding Clandestine Explosives," in *International Conference on Applications of Nuclear Techniques*, George Vourvopoulos and Themis Paradellis, eds., 2nd edition, Heraklio, Greece, 1990. [This provides an excellent review of the technologies.]

Porter, L., and D. A. Sparrow, "Assessment of Thermal Neutron Activation Applied to Surface and Near Surface Unexploded Ordnance," IDA Paper P-3339, 1997. [This provides a discussion of the phenomenology of soil and explosives, especially as it applies to slow neutron propagation. Data from field tests at two locations are analyzed. Models of signature as a function of ordnance size and depth are developed.]

Appendix P

# ELECTROCHEMICAL METHODS (PAPER I)

*Timothy M. Swager, Massachusetts Institute of Technology*[1]

## NEED TO DETECT A UNIQUE LANDMINE SIGNATURE

The detection of landmines has traditionally focused on secondary indicators that have given a high degree of false alarms. The classic technology that is still practiced is through the use of a metal detector. This was very effective in World War II, but, with the advent of landmines that are principally composed of plastic, the metal detector alarm needs to be set so low that for every 1 sq m of area checked, approximately 1.5 false positives are recorded. This has led to the extremely slow rate at which landmines can be removed, and a UN study reported that, at the present rate of removal, it will take over 1,000 years. In recognition of this problem, there has been a considerable effort put into such other detectors as ground-penetrating radar that alarm on density anomalies in the soil. The rationale for this approach is difficult to understand because soil generally has a heterogeneous composition and this method is unlikely to give a unique signal. Multiple other methods are being explored that seek

[1]AUTHOR NOTE: I have been assigned to detail electrical and optical approaches to landmine detection. However, to place my analysis in the proper context, I have felt it important to describe some of the basic issues of specificity and the need to alarm on a unique chemical signature of a landmine. As such, my first section may serve a general role in discussing the pros and cons of direct explosive detection, as opposed to conventional approaches that use secondary signatures. Although the latter approach has been pursued for many years and has consumed the vast majority of the funding, it seems unlikely to ever work.

to detect thermal anomalies associated with the disruption of the soil by a landmine. Density and thermal anomalies may assist in freshly buried landmines and, in these cases, soil disruption may be visually apparent; however, they cannot be considered as reliable signatures for all landmines. Such approaches are more a result of researchers reconfiguring existing technology than searching for the best solution. The saying comes to mind, "If all you have is a hammer, everything looks like a nail."

Focusing on secondary signals is flawed and gives a low degree of specificity and high false alarm rate. The continuing focus on these types of technologies is driven by the fact that the hardware exists, and they can have very fast response times. Although these technologies continue to advance slowly, with improved signal processing, new hardware, and fuzzy logic, the prospects for these technologies as an unambiguous indicator of landmines are low. The fact that this ground-penetrating radar only works in limited environments has led to restricted testing, and this practice was criticized by a recent General Accounting Office study [1].

The most reliable and rapid means to detect landmines is through trained canine-handler teams. Detailed research has shown that the dogs are indeed detecting the explosive vapor. As explosives are not naturally occurring materials, it is likely that they provide the most unique signature of a landmine. As a result, there has been considerable effort toward the detection of explosives in bulk or explosive vapors.

Two significant challenges face almost all chemical sensor technologies—sensitivity and selectivity. Landmine detection based on vapor methods presents daunting sensitivity requirements. The equilibrium vapor concentration of TNT, the principal explosive in landmines, is only 5–10 ppb. Vapor concentrations over a landmine, which are established by transport of the explosive compounds to the soil surface, are multiple orders of magnitude lower. (Tom Jenkins, whose work appears in Appendix Q, is an expert in this area.) DNT is a ubiquitous impurity in all explosives-grade TNT and because of its higher vapor pressure, detection of this analyte is also possible. Even higher sensitivity is necessary for rapid responses and the ability to sample a given suspected landmine location multiple times without depleting the small vapor headspace localized over the site. Specificity is achieved when the molecular (chemical or physical) properties of the explosive are an integral part of the transduction process.

In assessing sensor technologies for landmine detection based on detection of the explosive, it is important to realize that reports claiming sensitivity to specific concentrations of explosives can be misleading. The low volatility of explosives makes it easy to concentrate these materials and hence even a low-sensitivity sensor may be able to detect explosives if exposed long enough. There is a limited amount of explosive in the soil and surface over a landmine, and hence it is most proper to refer to detection limits in terms of the number of molecules or mass.

## ELECTRICAL TECHNOLOGIES

Detection principles based purely on electrical signals have advantages of being easily miniaturized and having low power requirements. The simplest method for detection is a change in the resistance of a transduction material. In general, semiconductive molecular solids and polymer films may have conductivities that are very sensitive to chemical vapors [2]. The general reason for this sensitivity is that the injected carriers in molecular and polymeric semiconductors are charged and electrostatic interactions of molecules result in trapping or untrapping of the carriers, thereby giving an increase or decrease in resistance. However, this inherent sensitivity also presents difficulties, and the sensor responses are generally variable based on the sample history, the morphology of the material, and the ionic impurities. More reliable sensors have been produced using inert intrinsically and electrically conducting particles of carbon dispersed in a swellable polymer [3]. The intrinsic low resistance of the particles causes them to be insensitive to electrostatic interactions. Molecular discrimination in this approach is the result of the relative solubilities of different molecules in the host polymer matrix. The specificity of individual sensors is very low and sensor selectivity is accomplished by constructing large arrays with a variety of host polymers. These types of sensor arrays are also known as artificial noses or electronic noses, and extensive measurements with different concentrations of analytes can be used to create algorithms that can be used to determine chemical compositions of complex orders. These technologies have been examined for landmine detection, and sub-ppb detection limits for DNT over 5-second exposures have been claimed [4]. However, this technology suffers from two general limitations. The first is that the polymers must swell with the explosive, and this requirement generally leads to low sensitivity. Although this technology may be improved (i.e., nano-structuring), the detection of landmines in any conventional operational concept will require many orders of magnitude greater sensitivity. A second limitation is the fact that arrays having elements of low specificity leads to overlaps in the response to different chemical signals. In controlled laboratory environments, the arrays can be made to function; however, outdoor field tests with a large variable humidity prove to be problematic. Higher orthogonality between the responses of all of the different sensors in the array can solve this problem, but such an advance is unlikely based purely on solubility differences. There has been considerable interest recently in organic Field Effect Transistors, and these materials have been suggested as having potential as sensors [5]. This approach has better prospects for sensitivity; however, it will still likely suffer from the same lack of sensitivity.

Electrochemical methods have also been examined for the detection of explosives. This approach differs from the other electrical methods in that the measurement involved chemical modification of the explosives or their degradation products. An advantage of this approach is that the chemical properties of the signal, namely the potential at which an electron is injected or removed, provide some specificity. Electrochemical sensors can function to give a response in potential (voltametric) or current (amperometric). A fundamental requirement of electrochemical methods is the need for a mobile electrolyte to maintain charge balance once an electron is removed or injected into the chemical being detected. Most often this leads to systems that operate in a solution phase and hence explosives must be extracted from soil or extracted from vapors. Nitroaromatics, TNT and DNT, have a high electron affinity and may be reduced directly. The low reduction potential (high electron affinity) of TNT has been reported to provide a selective electrochemical means to detect low ppb levels of TNT in water [6]. Similar claims of selectivity based on the unique reduction potential have been suggested in different electrochemical devices and electrode materials [7].

Secondary products from explosives, generally nitrogen oxides ($NO_X$), can also be detected electrochemically. This has been accomplished by thermally decomposing the explosives bound to soil and detecting the gaseous $NO_X$ products in amperometric sensors [8]. Nitro-groups can also be decomposed using the nitroreductase enzyme, and immobilizated films with a conducting polymer for mediation with the electrode can be used to produce amperometric sensors capable of the detection of two micromolar concentrations of TNT and DNT [9].

Based on the limited sensitivity of conventional electrochemical sensors it appears unlikely that this is a viable technology for the detection of landmines. Additional considerations that also limit this technology are the need for mobile electrolyte and the fact that electrodes can be easily fouled. Conducting polymer coatings can help in the latter regard; however, this also introduces more complexity to the sensor.

## OPTICAL DETECTION

Light of different wavelengths offers the chance to interrogate many different signatures that may be indicative of landmines. These range from infrared thermography to taking spectra of molecules.

Passive infrared imagining has been investigated in conjunction with other indirect methods in an attempt to increase selectivity through data fusion [10]. The interest in infrared solutions for landmine detection is likely related to the fact that there is an arsenal of Department of Defense knowledge in this area from the development of night-vision systems. The prospect that thermal anomalies over landmines, especially those that have been in the ground for extended periods, will be a robust indicator of landmines is extremely unlikely. Given the grave potential consequences of a system providing false negative readings, relying on this technique for detection rejection will never lead to an adequate discrimination.

There have been public claims of the ability of x-ray fluorescence for the detection of landmines. In this case the disclosed information has not held up to scientific scrutiny and appears to be principally intended to promote stock sales of a particular company [11].

In principle, spectroscopic techniques have the potential to provide the best selectivity for explosives. The infrared spectra of molecules can provide an information-rich fingerprint that allows for unambiguous identification. Direct detection by infrared absorption spectroscopy is not possible because of the limited sensitivity of this method. An alternative method with better sensitivity uses Surface Enhanced Resonance Raman Spectroscopy (SERRS), wherein light is scattered off an explosive molecule that is bound to a roughened highly polarizable silver or gold surface [12]. Although the SERRS method improves the sensitivity many orders of magnitude over conventional Raman Spectroscopy, this method still has insufficient sensitivity for landmine detection. In laboratory environments detection limits of $10^{-12}$ g of TNT were demonstrated and tests sponsored by the Defense Advanced Research Projects Agency systems could detect equilibrium vapor pressures of TNT (5 ppb) in 30 seconds. Hence, the sensitivity of this method is still multiple orders of magnitude too low.

Fluorescent methods are generally regarded as providing the highest sensitivity in conventional sensors. Hence there has been considerable interest in the detection of explosives by this technique. Fluorescence immunosensor displacement assays based on antibodies are widely used in biological sensing, and explosive detection systems have been developed on these principles. This method involves immobilization of TNT- or RDX-specific antibodies that are bound with a fluorescent label. Upon flowing a TNT or RDX solution over the antibody, the label is displaced by the explosive and a fluorescence signal is generated in the eluent phase [13]. Portable systems have demonstrated a 10 microgram per liter detection limit in water and this method has in principle very high selectivity and low detection limits. However, there are many factors that prevent this technology from being applicable to the landmine problem. The analysis in highly engineered systems [13] requires three minutes and prevents real-time landmine detection. Additionally, immunoassays have a number of limitations. The sensors must be kept at carefully regulated temperatures and are easily degraded or denatured in many chemical or thermal environments.

There have been multiple demonstrations of the direct detection of nitroaromatics by fluorescence quenching. The high electro-affinity of these compounds results in electron transfer from the excited states of fluorescent molecules. Sensors utilizing this principle have been developed using small molecule [14] and polymeric fluorophores [15]. In the small molecule case [15], array sensors have been developed with 0.2-second response times and are capable of low ppb detection limits in a few seconds. The array nature of this sensor provides superior selectivity over a single element sensor. Given that the electronic nature of the nitroaromatics is necessary for a response this system tends to have better discriminating ability than the electronic noses previously discussed that utilized differential solubility in polymer hosts.

The most versatile and sensitive explosive sensors to date are based on electronic polymers [15]. These systems have the ability to create gain internally because of the ability of these materials to transport their excited states from one spot to another [16]. The high mobility of the excited states in these extended electronic structures allows the excited state to visit a number of potential binding sites prior to emitting light. As a result of this mobility, there is a much higher probability that the excited state will encounter a nitroaromatic and lower detection limits result. Prototype landmine detectors have been developed with $10^{-15}$ g sensitivity [17], greater than a factor of 103 better than other systems. The materials used in this device have been shown to have size exclusion properties that also improve selectivity [18]. This method of using energy transport in electronically extended systems appears to be general, and recent results using porous luminescent silicon also showed good sensitivity [19].

An important feature of fluorescence-based detection methods is the ability to detect landmines at a distance. By spreading special fluorescent sensory materials over a minefield, it should be possible to image the location of all of the landmines. Initial results support the fact that this technology has good potential as a standoff sensor [20].

There are a few other methods for optical detection of explosives, such as photofragmentation with deep ultraviolet laser light followed with the fluorescence detection of nitric oxide [21]. This method gave a detection limit of 40 ppb of TNT in soil. Optical absorption has also been investigated by using a cavity to increase the pathlength of the light through the explosive vapor [22]. Competition assays using surface plasmon resonance have also been developed, but these assays are not robust and lack sensitivity [23].

## SUMMARY

The only robust primary indicator of a landmine is the explosive itself. There are many methods to detect explosives, however, most are limited by sensitivity and/or operational complexities. To date, only the fluorescence quenching method using amplifying electronic polymers has shown the necessary sensitivity in a portable device to detect landmines. This technique, in addition to allowing for handheld prototypes, can be used in standoff detection methods.

## REFERENCES

1. U.S. General Accounting Office, *Landmine Detection: DoD's Research Program Needs a Comprehensive Evaluation Strategy,* Report to the Chairman, Subcommittee on Military Research and Development, Committee on Armed Services, House of Representatives, April 2001.

2. L. L. Miller, J. S. Bankers, A. J. Schmidt, and D. C. Boyd, "Organic Vapors, Organic Polymers, and Electrical Conductivity," *J. of Phys. Org. Chem.,* No. 13, 2000, pp. 808–815.

3. M. C. Lonergan, E. J. Severin, B. J. Doleman, S. A. Beaber, R. H. Grubbs, and N. S. Lewis, "Array-Based Vapor Sensing Using Chemically Sensitive, Carbon Black-Polymer Resistors," *Chem. Mater.,* No. 8, 1996, pp. 2298–2312.

4. www.siam.org/meetings/sdm01/pdf/sdm01_11.pdf.

5. H. E. Katz, A. J. Lovinger, X. M. Hong, A. J. J. Dodabalapur, B-C. Wang, and K. Raghavachari, "Design of Organic Transistor Semiconductors for Logic Elements, Displays, and Sensors" in *Organic Field Effect Transistors,* F. Denis and B. Zhenan, eds., Seattle: Society for Optical Engineering, 2001, pp. 20–30.

6. J. Wang, R. K. Bhada, J. M. Lu, and D. MacDonald, "Remote Electrochemical Sensor for Monitoring TNT in Natural Waters," *Analytica Chemica Acta,* No. 361, 1998, pp. 85–91.

7. A. Hilmi and J. H. T. Luong, "Electrochemical Detectors Prepared by Electroless Deposition for Microfabricated Electrophoresis Chips," *Analytical Chemistry,* No. 72, 2000, pp. 4677–4682; J. Wang, F. Lu, D. MacDonald, J. M. Lu, M. E. S. Ozsoz, and K. R.

Rogers, "Screen-Printed Voltammetric Sensor for TNT," *Talanta,* No. 46, 1998, pp. 1405–1412.

8. W. J. Buttner, M. Bindlay, W. Vickers, W. M. Davis, E. R. Cespedes, S. Cooper, and J. W. Adams, "In Situ Detection of Trinitrotoluene and Other Nitrated Explosives in Soils," *Analytica Chemica Acta,* No. 341, 1997, pp. 63–71.
9. A. Naal, J. H. Park, S. Bernhard, J. P. Shapleigh, C. A. Batt, and H. D. Abruna, "Amerometric TNT Biosensor Based on the Oriented Immobilization of a Nitroreductase Maltose Binding Protein Fusion," *Analytical Chemistry,* No. 74, 2002, pp. 140–148.
10. www.onr.navy.mil/sci_tech/ocean/jcm/cimmd.htm.
11. www.stockblaster2000.com/LOCH.HTML and www.coralstrand.karoo.net/information/Extracts/ELF_1.4.htm.
12. C. J. McHugh, R. Keir, D. Graham, and W. E. Smith, "Selective Functionalisation of TNT for Sensitive Detection by SERRS," *Chemical Comm.*, 2002, pp. 580–581; K. Kneipp, Y. Wang, R. R. Dasari, M. S. Feld, B. D. Gilbert, J. Janni, and J. I. Steinfeld, "Near-Infrared Surface-Enhanced Raman Scattering of Trinitrotoluene on Colloidal Gold and Silver," *Spectrochemica Acta Part A—Molecular and Biomolecular Spectroscopy,* No. 51, 1995, pp. 2171–2175; J. M. Sylvia, J. A. Janni, J. D. Klein, and K. M. Spencer, "Surface-Enhanced Raman Detection of 2,4-Dinitrotoluene Impurity Vapor as a Marker to Locate Landmines," *Analytical Chemistry,* No. 72, 2000, pp. 5834–5840.
13. P. T. Charles, P. R. Gauger, C. H. Patterson, and A. W. Kusterbeck, "On-Site Immunoanalysis of Nitrate and Nitroaromatic Compounds in Groundwater," *Environmental Science and Technology,* No. 34, 2000, pp. 4641–4650.
14. J. V. Goodpaster and V. L. McGuffin, "Fluorescence Quenching as an Indirect Detection Method for Nitrated Explosives," *Analytical Chemistry,* No. 73, 2001, pp. 2004–2011; K. J. Albert and D. R. Walt, "High-Speed Fluorescence Detection of Explosive-Like Vapors," *Analytical Chemistry,* No. 72, 2000, pp. 1947–1955.
15. J-S. Yang and T. M. Swager, "Porous Shape Persistent Fluorescent Polymer Films: An Approach to TNT Sensory Materials," *J. Am. Chem. Soc.,* No. 120, 1998, pp. 5321–5322; Y. Liu, R. C. Mills,

J. M. Boncella, and K. S. Schanze, "Fluorescent Polyacetylene Thin Films Sensors for Nitroaromatics," *Langmuir,* No. 17, 2001, pp. 7452–7455; H. Sohn, R. M. Calhoun, M. J. Sailor, and W. C. Trogler, "Detection of TNT and Picric Acid on Surfaces and in Seawater by Using Photoluminescent Polysiloles," *Angewandte Chemie: International Edition,* No. 40, 2001, pp. 2104–2105.

16. T. M. Swager, "The Molecular Wire Approach to Sensory Signal Amplification," *Accts. Chem. Res.,* No. 31, 1998, pp. 201–207.

17. J. C. Cumming, C. Aker, M. Fisher, M. Fox, M. J. la Grone, D. Reust, M. G. Rockley, T. M. Swager, E. Towers, and V. Williams, "Using Novel Fluorescent Polymers as Sensory Materials for Above-Ground Sensing of Chemical Signature Compounds Emanating from Buried Landmines," *IEEE Transactions on Geoscience and Remote Sensing,* Vol. 39, No. 6, 2001, pp. 1119–1128.

18. Y-S. Yang and T. M. Swager, "Fluorescent Porous Polymer Films as TNT Chemosensors: Electronic and Structural Effects," *J. Am. Chem. Soc.,* No. 120, 1998, pp. 11864–11873.

19. S. Content, W. C. Trogler, and M. J. Sailor, "Detection of Nitrobenzene, DNT, and TNT Vapors by Quenching of Porous Silicon Photoluminescence," *Chemistry: A European Journal,* No. 6, 2000, pp. 2205–2213.

20. www.nomadics.com.

21. D. D. Wu, J. P. Singh, J. Y. Yueh, and D. L. Monts, "2,4,6-Trinitrotoluene Detection by Laser-Photofragmentation-Laser-Induced Fluorescence," *Applied Optics,* No. 35, 1996, pp. 3998–4003.

22. A. D. Usachev, T. S. Miller, J. P. Singh, F. Y. Yueh, P. R. Jang, and D. L. Monts, "Optical Properties of Gaseous 2,4,6-Trinitrotoluene in the Ultraviolet Region," *Applied Spectroscopy,* No. 55, 2001, pp. 125–129.

23. J. L. Elkind, D. I. Stimpson, A. A. Strong, D. U. Bartholomew, and J. L. Melendez, "Integrated Analytical Sensors: The Use of the TISPR-1 as a Biosensor," *Sensors and Actuators B: Chemical,* No. 54, 1999, pp. 182–190.

Appendix Q

# ELECTROCHEMICAL METHODS (PAPER II)

*Thomas F. Jenkins and Alan D. Hewitt, U.S. Army ERDC-CRREL*
*Thomas A. Ranney, Science and Technology Corporation*

## INTRODUCTION

Over the past several years, considerable resources have been expended to develop sensors capable of detecting buried landmines by sensing the vapors evolving through the soil and into the air above the mines. An effort was also made to characterize the vapor signatures of buried mines. This was done to provide the sensor developers with information on the qualitative and quantitative nature of the signature that is available for detection. Optimally, this second effort would have been completed first, providing specifications for the sensor community. This appendix addresses what is known about these signatures, their dependence on environmental constraints, and their implications for the probability of success of various chemical detection alternatives.

## QUALITATIVE NATURE OF THE SIGNATURE OF BURIED MINES

The initial work on the qualitative nature of the chemical signature of buried mines was conducted in the late 1960s and early 1970s with the sponsorship of the U.S. Army Mobility Equipment Research and Development Center, Fort Belvoir, Va. The instrumentation used at that time was primitive compared with what is available now, but the conclusions of these studies are relevant. With respect to explosives-derived signatures, laboratory tests indicated that equilibrium vapor concentrations of 2,4-DNT generally exceeded that of 2,4,6-TNT by at least an order of magnitude for military-grade TNTs [1]. 2,4-DNT is a manufacturing impurity in military-grade TNT and has been found to be present in most domestic and foreign TNTs [1,2]. For RDX (and Composition B), the most detectable vapor signature in laboratory tests was cyclohexanone, a solvent used in the manufacture of military-grade RDX [3]. Field tests indicated that cyclohexanone was detectable in the air above buried antitank landmines containing Composition B [4]. Composition B is composed of 60 percent RDX and 39 percent TNT.

In the more recent Defense Advanced Research Projects Agency (DARPA)–sponsored tests, chemical signature testing was limited to vapors that originated from military-grade explosives—largely TNT. Signatures attributable to other constituents of landmines, such as plastic casings, were not considered. Three samples of military-grade TNT were studied in depth: U.S. military-grade TNT obtained from Picatinny Arsenal, N.J., TNT taken from a Yugoslavian PMA-1A antipersonnel landmine, and TNT taken from a Yugoslavian PMA-2 antipersonnel landmine. Even though these TNTs were manufactured 35–55 years ago, 2,4-DNT and 1,3-DNB were the compounds detected at highest concentration in the vapor in equilibrium with the solid TNT [5]. 2,4,6-TNT was also detected in the vapor, but it was an order of magnitude lower in concentration.

The surfaces of four types of TNT-filled, Yugoslavian landmines (TMA-5, TMM1, PMA-1A, PMA-2) were sampled and in all cases the major signatures present were 2,4,6-TNT; 2,4-DNT; and 1,3-DNB. Often 2,4,6-TNT was the signature present at the highest concentration on the surfaces of these mines [6]. Similar results were found elsewhere [7]. These same mines were then buried at a research minefield at Fort Leonard Wood, Mo. Soil samples were collected and analyzed over the next three years to document the signature chemicals that accumulated in the surface and subsurface soils adjacent to these mines. Several thousand individual samples were collected and analyzed; the signature chemicals most often detected were 2,4-DNT; 2,4,6-TNT; and two environmental transformation compounds of 2,4,6-TNT, namely 2-amino-4,6-dinitrotoluene (2ADNT) and 4-amino-2,6-dinitrotoluene (4ADNT) [5,8].

Canadian researchers conducted a similar field study where seven different types of unfused landmines were buried in a gravel road. The soil was sampled three years after burial [9]. The same four signature chemicals (2,4-DNT; 2,4,6-TNT; 2ADNT; and 4ADNT) were found in these samples as found at the Fort Leonard Wood site.

Swedish scientists reported results from the analysis of soil samples collected in areas where various types of landmines had been excavated the previous week [10]. They reported the same suite of four signatures in soil samples from both a gravel road and deciduous forest. Soil samples were also analyzed from a Cambodian minefield. In these samples, only 2ADNT and 4ADNT were detected. From all these results, it appears that signature chemicals available in surface soils from TNT-filled landmines will largely be 2,4-DNT; 2,4,6-TNT; and its two transformation products, 2ADNT and 4ADNT.

## FREQUENCY OF DETECTION AND ESTIMATE OF SURFACE SOIL CONCENTRATIONS OF MAJOR SIGNATURE CHEMICALS IN SURFACE SOILS

Thousands of soil samples have been analyzed at the U.S. and Canadian research minefields to estimate the concentrations of the four major signature chemicals in surface soils above buried mines. It is the concentrations of these signature chemicals in the surface soils and their air/soil partition coefficients that control the levels of vapor signatures in the stagnant boundary layer air over a buried mine.

Results from the Fort Leonard Wood and Canadian Forces Base Suffield indicate that the frequency of detection and the concentrations of signature chemicals vary tremendously from one type of landmine to another [5,9]. For example, the frequency of detection of TNT-related signatures in surface soils for TMA-5 antitank mines (plastic cased) was much greater than that for a TMM1 antitank mine (metal cased), even though they contained equivalent masses of TNT. This appears to be ascribable to the TMM1 being better sealed than the TMA-5 mine. Similarly, the frequency of detection of the PMA-1A antipersonnel mines was much greater than for the PMA-2 mines that appear to be hermetically sealed. Even for the TMA-5 and PMA-1A, though, detection (concentrations greater than 1 ppb) for the four signature chemicals varied from 11 to 33 percent for the individual signature chemicals during four sampling periods over 15 months. For both types of mines, the frequency of detection increased in the order 2,4,6-TNT < 2,4-DNT < 2ADNT = 4ADNT. Median concentrations varied from about 44 and 17 ppb for 2ADNT and 4ADNT for surface soils over PMA-1A and TMA-5 mines, respectively, to 4 and 4 ppb for 2,4,6-TNT over PMA-1A and TMA-5 mines. Median concentrations of 2,4-DNT in surface soils were 32 and 16 ppb, respectively.

The relatively low concentrations of 2,4,6-TNT in these surface soils compared with the other signature chemicals appear to be attributable to a very short half-life for this compound. The half-life for TNT was estimated to be only about 1.1 days in Fort Leonard Wood soil at 22°C, whereas the half-life for 2,4-DNT was estimated at 26 days under identical conditions [11]. At Suffield, though, the mines were buried in a gravel road with apparently much less biological activity than at Fort Leonard Wood. The concentrations of 2ADNT and 4ADNT at Suffield were much lower relative to TNT than found at Fort Leonard Wood.

## CONCENTRATIONS OF EXPLOSIVES-RELATED SIGNATURES IN THE AIR ABOVE BURIED MINES

Estimates of the concentrations of TNT-derived signatures in the boundary layer air above buried mines have been made on the basis of surface soil concentrations of signature chemicals and their partition coefficients [5, 12]. A series of soil samples was collected near buried mines at Fort Leonard Wood. The soils were subsampled and a portion analyzed to estimate soil concentrations of 2,4-DNT and 2,4,6-TNT. The remaining soils were enclosed in glass vials, and the headspace was allowed to come to equilibrium. This headspace simulated the boundary layer air above buried mines. The headspace was sampled using solid phase microextraction and was then analyzed. These results together with the soil concentrations allowed calculation of soil/air partition coefficients, which varied somewhat from samples to sample but were generally about of $10^5$ for 2,4-DNT and $10^6$ for 2,4,6-TNT. When these values were combined with median surface soil concentrations obtained for soils above PMA 1-A and TMA-5 landmines, estimates for vapor concentration of 2,4-DNT and 2,4,6-TNT were about 200 pg/L and 1 pg/L, respectively [5]. If we assume that about 10 mL of boundary layer air is available for detection, the mass of signature available for vapor sensors would be about 2 pg for 2,4-DNT and 10 fg for 2,4,6-TNT. This assumes that all of the signature that is available could be collected. If we make an assumption that only about one-tenth of the available signature could be collected, then the mass of signature available would range from about 200 fg for 2,4-DNT to 1 fg for 2,4,6-TNT. Direct measurements of signatures in boundary layer air above buried mines at Fort Leonard Wood support these estimates.

Alternatively, if we could collect 1 g of surface soil particles for use in detection, the amount of signature chemicals available for detection would be about 20 ng for 2,4-DNT and 4 ng for TNT. This is an increase of $10^6$ in signature mass for 2,4-DNT and about $4 \times 10^7$ for 2,4,6-TNT. Clearly this tremendous increase would translate into a much more reliable detection system. It also allows the use of low vapor pressure compounds like 2ADNT and 4ADNT for detection.

In another study, the soil/air partitioning behavior was examined in laboratory experiments, with particular emphasis on the dependence on soil moisture content [12]. These experiments indicated that the range of equilibrium vapor concentrations can be reduced as much as 10,000 for very dry soils, making direct vapor detection more difficult in arid areas. These results explain the positive results that we have found for improving the detection capability of vapor sensors by the use of surface watering.

## RECOMMENDATIONS

The results discussed above, and the successful use of mine-sniffing dogs, clearly indicate that there is a chemical signature that can be used to detect the presence of buried mines. The source of the signature is surface soil contaminated from upward movement of signature chemicals. The nature of the explosives-related signatures has been characterized and is largely made up of four chemicals (2,4-DNT; 2,4,6-TNT; 2ADNT; and 4ADNT), the relative abundance of the four being dependent on the environmental conditions at the site. In some cases, for very poorly sealed landmines, the concentrations of these signatures in the surface soil can reach the ppm range, but more typically they are present in the low ppb range or below. While equilibrium between surface soil and boundary layer air is never really achieved, we can estimate the maximum concentration in the boundary layer air using an assumption of equilibrium. One must remember, though, that the volume of boundary layer air available for detection is very small, perhaps in the tens of milliliters at most. The equilibrium in the boundary layer air is controlled by the soil/air partition coefficient that is highly dependent on the moisture content of the surface soil. Even in the most optimum case, this equilibrium is at least $10^5$, indicating that the signature available for detection is orders of magnitude higher in the surface soil particles than in the air. Most attempts to develop chemical sensors for mine detection assume that sufficient signature would be available in the air above the soil to enable detection. We believe that the concentration of signature in the air above the mine is too low and too dependent on environmental factors to provide a reliable target for detection. Surface soil particles, however, have much more signature present and the availability of this signature is less environmentally dependent. While it is less convenient to engineer a detection scheme based on the presence of this signature source, a sensor system based on contaminated soil particles will provide a much more reliable basis for detection. A system of this type was configured several years ago, but the sensor utilized at the time was not adequate [13]. Combining this approach with a sensor based on fluorescent-quenching polymers [14] is a concept that should be considered for future development.

When sensors were tested during the DARPA Dog's Nose Program, it was assumed that there was a signature to detect when sampling directly above buried mines. Results obtained near the end of this program, however, revealed that the location and magnitude of the signature can be offset from the mine depending on the slope of the ground in the vicinity of the mine [15]. It appears that this is because the movement of the signature is dependent on soil moisture movement, which creates a downslope plume. In recent tests, we have utilized an onsite analytical technique to verify the location of the surface soil signature before conducting vapor sampling [15]. One team of sensor developers was on site with us and indicated that this capability would be very valuable when testing sensors to verify that there is a signature present where the sensor was tested.

## REFERENCES

1. D. C. Leggett, T. F. Jenkins, and R. P. Murrmann, *Composition of Vapors Evolved from Military TNT as Influenced by Temperature Solid Composition, Age, and Source,* U.S. Army Engineer Research and Development Center, Cold Regions Research and Engineering Laboratory, Special Report 77-16, 1977.

2. X. Zhao and J. Yinon, "Characterization of Origin Identification of 2,4,6-Trinitrotoluene Through Its By-Product Isomers by Liquid Chromatography-Atmospheric Pressure Ionization Mass Spectrometry," *Journal of Chromatography,* No. 946, 2002, pp. 125–132.

3. W. F. O'Reilly, T. F. Jenkins, R. P. Murrmann, D. C. Leggett, and R. Barrierra, *Exploratory Analysis of Vapor Impurities from TNT, RDX and Composition B,* U.S. Army Engineer Research and Development Center, Cold Regions Research and Engineering Laboratory, Special Report 194, 1973.

4. T. F. Jenkins, W. F. O'Reilly, R. P. Murrmann, and C. I. Collins, *Detection of Cyclohexanone in the Atmosphere Above Emplaced Antitank Mines,* U.S. Army Engineer Research and Development Center, Cold Regions Research and Engineering Laboratory, Special Report 203, 1974.

5. T. F. Jenkins, M. E. Walsh, P. H. Miyares, J. A. Kopczynski, T. A. Ranney, V. George, J. Pennington, and T. E. Berry, Jr., *Analysis of Explosives-Related Chemical Signatures in Soil Samples Collected Near Buried Land Mines,* U.S. Army Engineer Research and Development Center, Cold Regions Research and Engineering Laboratory, ERDC TR-00-5, 2000.

6. V. George, T. F. Jenkins, D. C. Leggett, J. H. Cragin, J. Phelan, J. Oxley, and J. Pennington, "Progress on Determining the Vapor Signature of a Buried Landmine," in *Detection and Remediation Technologies for Mines and Minelike Targets IV,* A. C. Dubey, J. F. Harvey, J. Broach, and R. E. Dugan, eds., Seattle: International Society for Optical Engineering, 1999, pp. 258–269.

7. E. Bender, A. Hogan, D. Leggett, G. Miskolczy, and S. MacDonald, "Surface Contamination by TNT," *Journal of Forensic Science,* 1992, pp. 1673–1678.

8. V. George, T. F. Jenkins, J. M. Phelan, D. C. Leggett, J. Oxley, S. W. Webb, P. H. Miyares, J. H. Cragin, J. Smith, and T. E. Berry, "Progress on Determining the Vapor Signature of a Buried Landmine," in *Detection and Remediation Technologies for Mines and Minelike Targets V*, A. C. Dubey, J. F. Harvey, J. Broach, and R. E. Dugan, eds., Seattle: International Society for Optical Engineering, 2000, pp. 590–601.

9. S. Desilets, N. Gagnon, T. F. Jenkins, and M. E. Walsh, *Residual Explosives in Soils Coming from Buried Landmines*, Defence Research Establishment Valcartier, Technical Report DREV TR-2000-125, 2001.

10. A. H. Kjellstrom and L. M. Sarholm, "Analysis of TNT and Related Compounds in Vapor and Solid Phase in Different Types of Soil," in *Detection and Remediation Technologies for Mines and Minelike Targets V*, A. C. Dubey, J. F. Harvey, J. Broach, and R. E. Dugan, eds., Seattle: International Society for Optical Engineering, 2000, pp. 496–503.

11. P. H. Miyares, and T. F. Jenkins, *Estimating the Half-Lives of Key Components of the Chemical Vapor Signature of Land Mines*, U.S. Army Engineer Research and Development Center, Cold Regions Research and Engineering Laboratory, ERDC/CRREL TR-00-17, 2000.

12. J. M. Phelan and J. L. Barnett, *Phase Partitioning of TNT and DNT in Soils*, Albuquerque, N.M.: Sandia National Laboratories, SAND2001-0310, 2001.

13. S. Desilets, L. V. Haley, and U. Thekkadath, "Trace Explosives Detection for Finding Landmines," in *Detection and Remediation Technologies for Mines and Minelike Targets III*, A. C. Dubey, J. F. Harvey, and J. Broach, eds., Seattle: International Society for Optical Engineering, 1998, pp. 441–452.

14. J. Yang and T. M. Swager, "Fluorescent Porous Polymer as TNT Chemosensors: Electronic and Structural Effects," *American Chemical Society*, No. 120, 1998, pp. 11864–11873.

15. A. D. Hewitt, T. F. Jenkins, and T. A. Ranney, *Field Gas Chromatography/Thermionic Detector for On-Site Determination of*

*Explosives in Soils,* U.S. Army Engineer Research and Development Center, Cold Regions Research and Engineering Laboratory, ERDC/CRREL TR-01-9, 2001.

Appendix R

# BIOLOGICAL SYSTEMS (PAPER I)

*Robert S. Burlage, University of Wisconsin*

## OVERVIEW

The best way to find buried ordnance (e.g., landmines) is to detect the explosive packaged inside. This expedient would eliminate detection of the ground clutter, such as shrapnel and stray metal fragments, that produce the great number of false positive signals and which slow down detection rates to unacceptable levels. The detection of the explosive is essentially what trained dogs do as they sniff out explosives vapors in the air above or near buried ordnance. While dogs remain the gold standard in landmine detection their application will always be severely limited. Another method must be found to identify explosive residue over wide areas.

The Microbial Mine Detection System (MMDS) is another example of a living system that responds to explosives and provides the operator with an identifiable signal. A common soil microorganism has been genetically engineered to recognize an explosive (DNT and TNT) and to respond to it by producing a fluorescent protein. These bacteria are sprayed over a field and allowed to contact the explosive that resides in the bulk phase of the soil. The concentrations of explosive are much higher here than in the vapor phase, and research has shown that ppm concentrations are available to the microbes—which is ideal. As fluorescent protein is produced, the bacteria become detectable using any of several fluorescence detection techniques. The fluorescent signals are mapped, and the area is examined for the source.

Although this technique has been demonstrated in the field, it has been starved for funding and never developed in the appropriate manner. Significant problems remain but should be solvable. Thus it remains a largely untested and theoretical approach to broad area detection. The technique itself has been patented [1].

## STATE OF DEVELOPMENT

As mentioned above, this technology should be considered basic laboratory research. The technology involving recombinant bacterial strains has been performed by the principal investigator for many years. Detection of fluorescence has been practiced far longer. The application of such bacterial strains in an environmental setting and the rapid detection of fluorescent signals over large areas are relatively new concepts, and much work remains before they can be considered optimized.

However, the technology has been field tested. This test was somewhat of a rush to the field, and the principal investigator wished for more time to adequately test the technique, but one must take opportunities as they are presented. The results of this test are available [2]. Important details of the technique and the field test have also been reported [3]. During this test the MMDS technique detected all five of five targets within a quarter-acre site, within a distance of 1 m. Targets ranged in size from 4 oz to 10 lb of explosive. There were two false positive signals, defined as being outside the 1-m radius from a target. However, it was hypothesized that these signals were not false at all, but rather they were small fragments of TNT that were physically transported away from the target by water flow or animal activity. In the absence of a definitive assay, this hypothesis remains unproved but highly likely. The test required several hours for detection, although it should be emphasized that this was the first field test and much time was taken in deciphering unique signals. As more is known about the capabilities in the field, this analysis time will greatly decrease. Our goal is to cover a 10-m-wide strip of ground at walking pace (3 miles per hour) or slightly faster.

The MMDS technique has been tested in two other locales. The first was at a U.S. Air Force base in the desert and the other at a closed ammunition plant in an eastern deciduous forest. Both tests were conducted under suboptimal conditions. In the first test, a charge of

TNT was detonated and the area downwind of the site was examined for dispersed fragments. In the second test, the bacteria were sprayed over an area known from historical records to be contaminated with a number of different explosives. Positive signals were found in each case, although the tests did not utilize discrete point sources, and therefore a detection rate cannot be produced. Reports on these two releases are available from the principal investigator.

It is clear from these few tests that the system should function under a certain set of parameters. Once applied, the bacteria require about four to six hours before they maximize their output of fluorescent protein (and thus become optimally detectable). About $10^6$ bacteria per square centimeter are preferred. The signal is then stable for the next 24 hours, although the bacteria die off exponentially over the next several days. Thus the objective in mine clearance is to spray relatively large areas (hundreds of acres) at a time; the slowness of the incubation period is balanced by the large area covered. The defining factor in speed then becomes the fluorescence detection step.

## STRENGTHS

This technique has several unique attributes. Bacteria can be grown easily and without great expense: They need only a solution of sugar and some inexpensive chemicals. The cost of the technique is outlined below but can be considered very affordable—even for developing nations. In addition, the technique costs the least when very large areas are examined. This contrasts with many other techniques that have a set cost per area covered.

The system is dependent on a bacterial regulatory protein that recognizes the shape of the explosive chemical. Only structurally similar molecules have this same "lock and key" effect, and the mere presence of nitrogenous sources in the soil is insufficient to trigger the effect. It is likely that other bacterial strains can be produced which are responsive to other explosives, such as RDX and HMX. However, these have not yet been produced. This is a matter of scientific inquiry and requires funding, but there is no good reason why a suitable strain cannot be produced. The various strains can then be mixed together in a cocktail of explosive-sensitive bacteria.

Many present and proposed techniques for mine clearance are adversely affected by the presence of vegetation. The MMDS technique is actually improved by the presence of vegetation, which appears to conduct the explosives and magnify the signal. MMDS may be the only suitable system for many areas of the world where removal of vegetation would have adverse environmental consequences.

Because the technique detects the presence of the explosive chemicals, and not the ordnance package, it will not be affected by stray bits of metal. While it may find trace amounts of chemical residue from recent explosions, it will probably find only those signals that result from a continuous source of explosive at low concentration, such as a buried landmine. It will also find a raw explosive that is buried or find discarded ammunition that has been forgotten. It has been suggested that this system may work best as one of several complementary techniques for the same field.

The MMDS system is envisioned as a fully portable system. The major weight consideration is the wet weight of the bacteria. At present, we transport the bacteria as a thick paste, which stays fresh for a couple of days. However, we have some experience with freeze-drying bacteria to remove the water weight and increase their longevity. It is also possible to grow the bacteria on site using a portable fermentor. This would make the technique portable to any area of the world. Based on the final configuration of the fluorescence detection system, the initial capital cost might be a little high. However, this is a onetime charge (not considering spare parts) and the remainder of the system then becomes very affordable. There is also little cleanup afterward, as the bacteria die off rather quickly and actually serve as a fertilizer for the soil.

## WEAKNESSES

There are still major weaknesses of the system that must be addressed, as befits a basic research project. Our field trials have suggested many of the problems. For example, we still need to know much more about the partitioning of explosive chemicals between soil (clay, silt, sand, organic) and the bulk phase because only the bulk phase material is ultimately detectable. We have also had substantial problems with dispersal of the bacteria over dry ground, where the bacteria will be quickly absorbed by the soil and the signal lost to detection. We have experimented with encapsulation techniques to overcome this difficulty, but there is a substantial tradeoff in effectiveness. Other techniques are possible.

To work effectively, the bacteria must physically contact the explosive. In addition, the overall soil conditions must not be too deleterious, considering that a living system must be used. Therefore, a snowy field would be too cold for application and would block the bacteria from the explosive in the soil. Recent rain would change the location of the signal, as we have seen in the field. Of course, this limitation is common to other methods that detect the explosive itself.

It has been suggested that the use of a microorganism for mine clearance would constitute the use of a biological weapon, as defined by international weapon control treaties. An objective reading of these treaties indicates that the use of MMDS in a battlefield scenario would undoubtedly violate the treaty, although use in humanitarian demining, even broadly defined, would not violate the treaties at all.

It has also been suggested that this technique would not receive widespread approval because it utilizes genetically engineered microorganisms. This has not proved to be the case. We work closely with the U.S. Environmental Protection Agency on the release of recombinant bacteria and only with its approval. All three field releases have been performed without incident. Other countries have expressed interest in the technology even after the components of the system are explained to them. There seems to be no great obstacle due to the nature of the bacteria.

Finally, the fluorescence detection system has, in the past, proved unable to clearly identify a signal in many instances. To a great degree, this is a learning activity as we decide which fluorescent intensity is background and which is above the threshold for a positive signal. Clearly, alterations in the detection system and possibly in the fluorescence protein used will have to be considered. Because the ultimate goal is the production of an airborne system that can cover hundreds of acres per day, the detection system will have to meet size and weight requirements.

## PROBABLE COST OF A FUNCTIONAL SYSTEM

It has been estimated that most of the difficult questions regarding the viability of this system could be answered in about two years of research at a total cost of $650,000. This research would address critical questions about the stability of the genetic construction, partitioning of explosive residue in various soil types, the role of vegetation in uptake and magnification of the signal, and reliable dispersal methods that allow the system to be used in many soil types. In conjunction with these tasks, the application of commercial fluorescence detection systems should be evaluated to determine whether "off the shelf" technology could be successfully employed in an airborne system.

Final costs are quite speculative at such an early phase of the project. Major capital equipment costs will be required for an on-site fermentor (if needed) and for the fluorescence detection system. I estimate these costs to be in the $250,000–300,000 range. Subcontracting the fermentation process is a better option and will save significant sums. In addition, the airborne platform (probably a helicopter) will also need a subcontract. Dispersal of the bacteria is relatively inexpensive—mostly unskilled labor. The goal is to cover an acre of ground in small patches for $10–40, while the price would drop below $10 per acre for larger areas (hundreds of acres).

## POTENTIAL IN THE SHORT TERM

As noted above, a two-year basic research program would answer most, if not all, of the questions that funding agencies have raised after our earlier tests. This would be a team approach because research is needed in several areas: microbiology, botany, engineering, fluidics, etc. After this time it should be ready for defined-field testing or for real-world conditions. I would like to analyze an actual minefield to map the signals and then have the field cleared using conventional technologies to determine our efficacy as well as our nearness to target and any confounding influences. While defined-field tests can do this to some extent, an actual minefield to gather data is best. Because our technique is nondestructive of the field, the mine clearers will not be influenced by our results. I am particularly interested in any mines that we may miss. Will they contain a novel explosive? Are they associated with a certain soil type (such as high clay content)? Were they placed in the ground recently? In our first test, we used targets that had been buried for three months—but real-world conditions may not be similar. We have contacted many interested individuals in other countries with access to minefields. They are reticent about funding the project because too many unknowns remain. We would like to approach them again with an improved and more predictable system.

With a reliable system, a best-case scenario of 100-percent detection with essentially no false positive signals can be predicted. This is a bold prediction, but it is based on the specificity of the explosive for the bacterial regulatory protein, the sensitivity of the system compared with known leakage concentrations, and the demonstrated sensitivity of fluorescence detection systems (which are assumed to improve even more). The typical sites for these predictions would be old (1–20 years) minefields, such as those found in Bosnia, many parts of Africa, the Falkland Islands, and many parts of Southeast Asia. Efficiency of the system may suffer in areas where the environment is more hostile, such as in desert regions (e.g., the Middle East). However, encapsulation techniques for the bacteria and judicious timing to avoid the extreme temperatures may dramatically improve the technique. MMDS is not predicted to work well in those areas with standing water, consistently low or high temperatures, or snow-covered soil.

## REFERENCES

1. R. S. Burlage, K. Everman, and D. Patek, *Method for Detection of Buried Explosives Using a Biosensor,* U.S. Patent No. 5,972,638, 1999.

2. W. Schaefer, *Test Report for the Microbial Mine Detection System (MMDS),* DSWA IACRO HD1102-8-1490-097, 1998.

3. R. Fischer, R. Burlage, J. DiBenedetto, and M. Maston, "UXO and Mine Detection Using Laser Induced Fluorescence Imagery and Genetically Engineered Microbes," *Army AL&T,* July–August 2000, pp. 10–12.

Appendix S

# BIOLOGICAL SYSTEMS (PAPER II)

*Jerry J. Bromenshenk, Colin B. Henderson, and Garon C. Smith,*
*University of Montana*

## OVERVIEW

Honeybees offer the potential of using free-flying organisms to search wide areas for the presence of explosives, unexploded ordnance (UXO), and landmines. The use of bees is analogous to dogs for mine clearance, except that a colony of tens of thousands of bees can be trained in about one hour to fly over and search a field for explosives, does not require a leash, and will not set off any mines. Like dogs, bees can be trained to search for either the odors of individual explosives or suites of these chemicals.

Initial tests indicate that bees are capable of detecting these odors at concentrations below those detectable by most instruments. Bees approach, if not match, the odor sensitivity of dogs, i.e., low ppt to ppq, and possibly lower. Field trials with honeybees at ppb and ppt vapor concentrations of 2,4-DNT (a residue in military-grade TNT) showed a detection probability of 97–99 percent. We calculated a 1.0–2.5 percent probability of false positive, and less than 1 percent probability of false negative, based on three different statistical sampling strategies [1,2]. Calculated receiver operating characteristic (ROC) curves for 10 ppb through 0.001 ppb indicated that for doses higher than 0.01 ppb (10 ppt), the bee system behaves like a very fine-tuned, nearly ideal, detector. In addition, bees have mop-like, electrostatically charged hairs that enable them to bring back to their colonies samples of explosive chemicals as well as biological agents and other harmful materials [3,4,5].

## PHYSICAL PRINCIPLES AND MINE FEATURES EXPLOITED FOR AREA REDUCTION

Like dogs, the use of bees exploits the ability of an organism to detect landmine chemical signatures at very low concentrations. Unlike most instruments, bees and dogs can be trained to search for either specific chemicals or suites of chemicals. However, dogs are usually kept on a leash, and most chemical-sensing instruments have to be carried across minefields. Because free-flying bees search outward from their hives and then return to their hives, fairly large areas can be searched and samples recovered without risk to humans. In addition, bees continually cross and recross forage areas, ranging as far as 1–2 km. Consequently, bees are likely to encounter explosive vapor plumes multiple times, amplifying low concentrations. In other words, the bees autonomously and intensively search each region.

## STATE OF DEVELOPMENT OF BEE LANDMINE FINDER

Under the leadership of the University of Montana (UM), a consortium of Defense Advanced Research Projects Agency (DARPA)–sponsored investigators has been exploring ways to leverage the natural foraging behavior of honeybees for military applications, including detection and localization of explosive materials and landmines.

Over the past three decades, demonstrations by UM at U.S. Environmental Protection Agency Superfund sites, Department of Energy reservations, and Department of Defense installations have shown that honeybees effectively gather chemical information over ranges extending several kilometers. Honeybees have been used to monitor trace elements and heavy metals, volatile and semivolatile organic chemicals, radioactive materials, and (more recently) explosives, in both the United States and Europe. As honeybees collect water and forage for nectar and pollen, the electrostatic charge on their body attracts dust, pollen, soil, and other particles, including chemicals leaking from explosive devices into soil and water. These materials are then returned to the hive. These capabilities are summarized in a newly released book entitled *Honey Bees: Estimating the Environmental Impact of Chemicals* [6].

The U.S. Army Center for Environmental Health Research contracted with UM to complete a seven-year series of applications of honeybees for surveying and monitoring industrial and militarily unique chemicals at Aberdeen Proving Ground (APG), Md. These investigations pioneered examination of volatile and semivolatile chemicals (including explosives) in the air inside beehives and developed electronic hives capable of continuously monitoring colony dynamics and of controlling chemical sampling equipment associated with the hives. The objective was to trigger a sampling event when critical bee performance traits displayed a significant change, such as altered forager flights, reduced numbers of bees returning to the hive, or a breakdown of the complex cooperative behaviors, such as maintaining constant temperature in the brood nest. Also, all this information was ported from remote field sites via communication links to a central facility in Montana.

A DARPA program that began in 1999 brought together a team of scientists from UM, Sandia National Laboratories (SNL), Oak Ridge National Laboratories (ORNL), and Southwest Research Institute (SwRI) to focus on developing bees for use in detecting chemical and biological warfare agents as well as landmines. Partners at the Air Force Research Laboratory (AFRL) and the Air Force Force Protection Laboratory joined the research team in 2000.

As a result of work by UM, SNL, and ORNL, accompanied by field trials at APG, honeybees, pollen, wax, honey, and the air inside the hive can now be analyzed for explosives. Analytical technologies including Solid Phase Microextraction fibers (SNL), and a new generation of sorbent sol-gels (ORNL), combined with gas chromatography and mass spectrometry have been field validated for detection of explosives. Tested compounds include nitroglycerine (NG); diethyleneglycoldinitrate (DEGN); 2,6-DNT; 2,4-DNT; TNT; RDX; and pentaerytnitoltetranitrate (PETN) in bees and their hives [7,8].

## BEES TRAINED TO FIND TRACE LEVELS OF EXPLOSIVES

In addition to this passive collection of materials by untrained bees, more recent field trials have shown that honeybees can also be trained to actively search for and detect explosive materials. The training technologies were developed by UM. During field trials at SwRI in 2001, sand-explosive mixtures on the ground provided point-source targets and generated plume vapor concentrations of 2,4-DNT in the 0.7–13.0 ppb range. Honeybee flights over blanks, sand controls, and the targets were continuously monitored with video cameras and analyzed off line. Using detection algorithms designed for radar applications (AFRL), honeybees were shown to locate the DNT within minutes of when the colony began to forage. The false positive rate was 2.5 percent, while the false negative rate was less than 1 percent. Concurrent statistical data analysis of cumulative counts of bees over blanks, controls, and targets by UM and SwRI yielded a detection probability of 98.7 percent, with less than 1 percent false positive and 1 percent false negative results. Continuing 2002 field trials have pushed thresholds of detection an order of magnitude lower, with bees reliably detecting targets in the low ppt range and well within the range of measured vapor concentrations above active landmine fields. *An informal comparison at SwRI of the Nomadics Fido sensor, under identical field conditions showed equal performance by Fido and by bees in detecting 2,4-DNT vapor plumes—with the threshold of detection virtually the same for the instrument and for the bees. Bees correctly located and identified DNT targets in an area 200 m across in less than one hour with no need for a person to enter the test area. For Fido, the operator had to be led to the target so that it could be "sniffed." Searching the 200-m-diameter test area for a small DNT target would have required hours or days.*

This initial bee project, which began as an unproven concept, has in three years developed into a system for which preliminary performance capabilities and operating characteristics can be quantified. Overall, the costs of the combined bee testing have averaged about $1 million per year. This includes laboratory assessment of odor detection, greenhouse flight experiments, development of training protocols for bees at a one-acre tent and under open field conditions at SwRI, and improved analytical methods and devices for measuring explosive residues and vapors in bee systems.

Because of the encouraging results from recent field trials, UM proposed additional technology demonstrations to DARPA for spring 2002. These included detection of UXO at weapon ranges, tracking of honeybees with Lidar or digital imagery, and detection of buried explosives and actual landmines at landmine test beds.

A primary goal is to determine minimum detection limits for honeybees, comparing those to minimum detection levels exhibited by trained dog teams. If successful, honeybees would provide an economical and effective method for wide area monitoring and localization of UXO and landmines. *As mentioned, preliminary results from the new series of audited field trials at SwRI have shown that bees can easily locate, in less than an hour, a 2,4-DNT target placed approximately 100 m from the hive. The targets emitted trace vapor plumes measured in the low ppt range.*

## CURRENT CAPABILITIES AND OPERATING CHARACTERISTICS

At UM, using a benchtop bioassay called the Proboscis Extension Response System, individual bees were trained to discriminate between a small puff of air containing no scent and a puff containing an explosive's odor. The bee proved to be able to reliably detect (p less than 0.001) the presence of 2,4-DNT at 20 ppb. Experiments conducted by ORNL revealed that surface chemistry interactions in glass tubes of the dose-delivery system—not the bees' sensitivity to odor—limited the level to which the bees could respond. As such, the minimum detection limits of bees could not be established using this system. Follow-up studies with free-flying bees in a greenhouse again showed that bees could detect a 10-percent mixture of DNT in soil and would search for and hover over DNT-spiked targets. The bees' ability to discriminate between blanks and DNT was almost perfect (p less than 0.001).

In summer 2002, UM researchers supported by SwRI scientists completed a series of double-blind, fully audited trials. UM trained the bees and SNL explosives experts worked with SwRI chemists to estimate and then measure the dose delivered at each target. UM and SwRI statisticians analyzed the cumulative count data, while AFRL digitized and subsampled the video data and then calculated the probability of detection (PD) and approximate ROC.

The ROC curve shown in Figure S.1 is the relationship between the PD by the receiver when the target or agent is present and the probability that the device or receiver will register the target or agent present, when it is in fact absent [9]. In biological terms, the ROC curve relates the probability of a true positive to the probability of a false positive. The ROC curve presented is for honeybees trained to

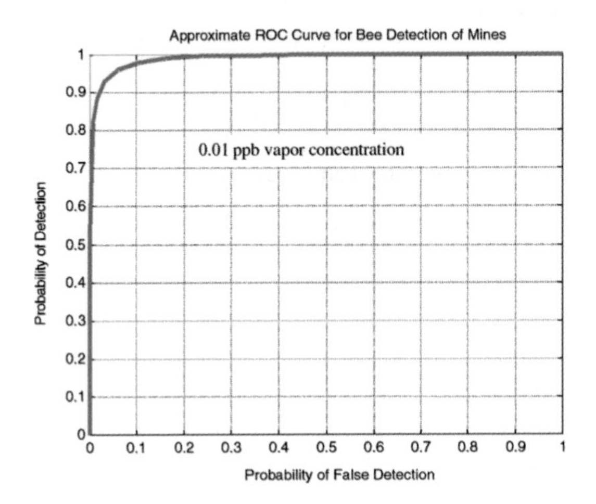


**Figure S.1—ROC Curve for Honeybees Trained to Associate the Smell of DNT with Food**

associate the smell of DNT (the most common signature of explosives in landmines) with food. The diagram plots 0.01 ppb lower level in the vapor plume above DNT placed under dry sand in petri dishes. The ROC curves for these concentrations are very near 1.0 and 0, suggesting that at these vapor concentrations, honeybees perform as virtually an ideal detector.

Bees had little difficulty detecting 0.7–13.0 ppb DNT vapor in the controlled experiment at SwRI in 2001. The 2001 levels of DNT are likely to be higher than those found above most buried landmines, but the experiment was designed to characterize the potential use of bees for landmine or explosive detection and not to simulate use of honeybees in mine removal operations. Subsequent trials in Montana reduced the amount of DNT to 0.5 g of explosive in 300 g of sand, with an estimated vapor concentration from 50–70 ppt. There was little change in the bees' ability to detect the target out to 100 m from the hive (DNT detection, p less than 0.0001 at an observed power of 1). The curve in the ROC diagram above is calculated from these Montana trials. Although these results represent two-hour trial periods, data subsampling showed high statistical significance from as little as four minutes of data. The audited trials of low DNT concentrations recently concluded at SwRI confirm our findings in Montana. Targets containing 0.1, 0.5, and 1.0 g of DNT were evaluated, and plume concentrations from 20 to 300 ppt were measured. Bees consistently detected targets generating 50–80 ppt vapor and under moist conditions detected plumes down to 30 ppt.

The experimental conditions for the SwRI trials were optimized for good bee performance. Ground cover was limited and wind conditions were mild. Terrain was flat, and there were few or no flowering plants to divert the attention of the bees from the target. Operating temperature was high, thereby increasing vapor pressure above the target. However, the sand and DNT were dry—conditions that tend to decrease probability of detection by dogs and many instruments. In Montana, the temperatures were cooler, and the vapor pressure was much lower. Based on actual plume measurements combined with models of the effects of wind, DNT concentrations were estimated by Sandia explosives experts to be in the low ppt, and possibly ppq, depending on how far from the target the bees were detecting the leading edge of the vapor plume.

The ROC calculations should not be interpreted to mean that honeybees are perfect detectors of landmines in real-world conditions. However, these structured and audited field trials demonstrated significant detection capability.

## LIMITATIONS

Bee detection trials using real landmines have yet to be performed. Bees do not fly at night, in heavy rain, or in cold weather (below 40°F). The effects of such environmental conditions as dry versus wet soil, open areas versus dense forests, different climatic conditions, and interference or complicating factors (e.g., multiple targets, mine fragments, exploded ordinance, chemical smokes) have not been examined. Bees perform well in open fields under hot and dry conditions. Whether they would perform as well in dense forests is unknown. The principal factors limiting performance are lack of knowledge about the search patterns of bees in different settings and the inability to track bee movements.

## POTENTIAL FOR IMPROVEMENT (TWO TO SEVEN YEARS)

Many of the aforementioned complicating factors can be tested and evaluated within two years. Practical experience and testing at real minefields is essential. Conditioning protocols need to be fine tuned to accomplish signal amplification (i.e., more bees over a target—estimated one-year effort). However, for use in demining applications, a tracking system that can locate and follow bees, either by their own chemical signatures, explosives adsorbed onto the bees, or some form of an inexpensive marker, such as a reflective powder, is needed. Systems to be tested range from high-resolution cameras that could be flown over the site or to those used to look down from satellites to Lidar, which could discriminate bees by chemical signatures 1–2 km away from their hive. The tracking problem should be solvable in two to three years, maybe five at the most. Assuming that bees can detect real landmines without a serious loss in the PD and in the ROCs, it is conceivable that the bee system might provide a near ideal detector. Combined with an imaging or tracking system capable of discriminating bees over targets (e.g., landmines) and the software to automate the process, a Lidar or camera system combined with bees, in the best of conditions, should be able to survey a 1–2 sq km area within a few hours.

There may be utility in also developing a decidedly low-technology counterpart. In theory, it should be feasible to teach local beekeepers to train their own bee colonies to survey fields as part of an area reduction program for the price of some syrup, some training standards, a ladder, and a pair of binoculars. Even this crude setup could track and count bees at distances equal to or farther than the leash currently used on minefield dogs.

## SENSIBLE RESEARCH AND DEVELOPMENT PROGRAM LEADING TO A FIELD-USABLE DEMINING SYSTEM WITHIN 24 MONTHS

Surveying areas by examining chemical residues brought back to the hive by bees is based on 30 years of proven technology. In less than four years, bees have been shown to be trainable and to detect explosives vapors at very low concentrations with a very high probability of detection and a very low probability of false negatives. Developing this system into a usable tool for landmine detection, especially area reduction, should be possible within two to five years at a cost of $1 million per year, based on previous research.

Critical research needs include

- Examining the effects of environmental factors, multiple targets, and interferents on odor discrimination and PD with respect to wide area search. Anecdotal evidence from the 2002 SwRI trials indicate that simply wetting the soil may dramatically improve bee detection of minute amounts of explosives. Also, bees can be kept continually conditioned and able to find trace amounts of explosives for more than 36 days, despite changing climatic conditions ranging from a drought to a 500-year flood.
- Determining the limits of detection (Are they really as good as the dog, or are they just better than most instruments?):

  —Test under controlled conditions (e.g., dry soil versus wet soil).

  —Test at real minefields in different parts of the world.

  —Compare competing technologies, such as Nomadics' Fido (informal trials, 2002 demonstrated similar detection thresholds for Fido and bees).
- Defining how bees search and their movement patterns in different habitats, settings, and climates:

  —Use Lidar, radar, video imaging, or taggants and pigments to establish foraging patterns and bee densities.

  —Assess residue/breakdown chemistry of explosives and chemical agents for other promising training odors.

  —Optimize/enhance at-the-hive analytical protocols for quick identification of agents in the area to be searched.
- Developing improved methods for tracking/spotting of bees as they search and when they are over targets (e.g., landmines), including

  —Technology improvements for the military to provide order of magnitude improvements in the speed at which antipersonnel mines can be cleared (e.g., Lidar or high-resolution cameras mounted on aircraft or satellites).

—Low-cost approach enabling local beekeepers to help effect area reduction (e.g., bees, binoculars, ladders). On average, there is a colony of bees per every 0.5 km around the world.

Finally, the potential use of bees for humanitarian demining has begun to result in inquiries from demining centers and companies situated around the world. To date, no other system offers the potential of standoff detection and surveying of areas ranging from 100 m to 1 km or more across. All lines of evidence indicate that bees should be able to find real landmines (or at least, those that leak trace amounts of explosives). The missing aspects are experience on real minefields, so that the system can be optimized and thoroughly calibrated, and a tracking system, such as a portable Lidar, that would provide standoff tracking of bees so the minefields could be mapped.

## REFERENCES

1. P. J. Rodacy, S. F. A. Bender, J. J. Bromenshenk, C. B. Henderson, and G. Bender, "The Training and Deployment of Honeybees to Detect Explosives and Other Agents of Harm," Proceedings of SPIE AeroSense CBS Work, April 2002.

2. M. Block, R. Medina, and R. Albanese, *Analysis of Data Determining Whether European Honey Bees Can Detect DN*, Air Force Research Laboratory, Technical Report to DARPA (in progress).

3. B. Lighthart, K. R. Prier, G. M. Loper, and J. J. Bromenshenk, "Bees Scavenge Airborne Bacteria," *Microb. Ecol.*, No. 39, 2000, pp. 314–321.

4. G. C. Smith, J. J. Bromenshenk, D. C. Jones, and G. H. Alnasser, "Volatile and Semi-Volatile Organic Compounds in Beehive Atmospheres," in *Honey Bees: Estimating the Environmental Impact of Chemicals*, J. Devillers and M. Pham-Delegue, eds., New York: Taylor and Francis, 2002, pp. 12–41.

5. D. Barasic, J. J. Bromenshenk, N. Kezic, and A. Vertacnik, "The Role of Honey Bees in Environmental Monitoring in Croatia," in *Honey Bees: Estimating the Environmental Impact of Chemicals*, J. Devillers and M. Pham-Delegue, eds., New York: Taylor and Francis, 2002, pp. 160–185.

6. J. Devillers and M. Pham-Delegue, eds., *Honey Bees: Estimating the Environmental Impact of Chemicals*, New York and London: Taylor and Francis, 2002.

7. P. Rodacy, Sandia Laboratories, Memorandum to DARPA, October 2001.

8. J. J. Bromenshenk, "Engineered Bee Colonies: A Platform for Bioreporting and Seeking Agents of Harm," Program Review, Defense Advance Research Projects Agency, Controlled Biological Systems and Biomimetic Review, Breckenridge, Colo., 2001.

9. H. L. Van Trees, *Detection, Estimation, and Modulation Theory,* New York: John Wiley & Sons, 1968.

Appendix T

# CANINE-ASSISTED DETECTION

*Håvard Bach, GICHD*
*James Phelan, Sandia National Laboratories*

## HUMANITARIAN LANDMINE DETECTION NEEDS

Buried landmines are difficult to find because they are designed for concealment. These small, low-tech devices appear with a wide variety of designs and materials of construction, and can barely be differentiated from stones, roots, and scrap material in the ground. Yet, they always contain an explosive—most typically, TNT. The simplicity of landmines is thus the greatest challenge for those researchers attempting to develop new or improved methods for detection.

Detection of landmines is, however, only part of the problem. We know that only a small percentage of land typically contains mines. Up to 90 percent of mine-suspected areas could be released and given back to local societies if we knew where the landmines were *not* buried. The humanitarian demining challenge is therefore two-fold:

- Detect the areas that are free from landmines for immediate release (area reduction).
- Determine the exact location of landmines, enabling removal or destruction.

New or improved technology must also satisfy most or all of the following criteria to improve humanitarian demining:

- Help accelerate the demining process.
- Be as safe or safer than existing technology and approaches.

- Be practical to use and easy to repair and maintain.
- Be affordable.
- Enhance overall cost efficiency of demining.
- Not be too complicated for use by deminers.

Demining is a complex multitask process. There are great variations in types of landmines, landscape, terrain, vegetation, soil properties, weather conditions, burial depths, and methods for deploying landmines. Demining further involves many activities: survey, minefield boundary location, removal of vegetation, tripwire detection, ground preparation, and pinpointing and removal/destruction of mines and unexploded ordnance. No single tool can address all these tasks effectively. Thus it is necessary to apply a process with complementary tools and techniques.

The historical approach for landmine detection has been manual mine clearance. While this technique is highly reliable for clearance, it is unfortunately slow and dangerous. Metal detectors used with manual clearance methods are unsuitable for an entire class of low-metal-content mines. Mechanical mine clearance has evolved from the use of military tanks with flails or rollers to commercially produced machines with higher degrees of mobility and reliability. Technology evolution has been slow and mechanical clearance continues to have limited potential in many areas. Trace chemical detection—using dogs—is a versatile tool; yet, much more work is needed to understand optimal applications in the field. Humanitarian demining continues to need improved methods in all activities and research can bring the needed improvements in many areas, especially trace chemical detection.

## PHYSICAL PRINCIPLES OF TRACE CHEMICAL DETECTION

The nature of the landmine chemical signature is a complex phenomenon that is now reasonably well understood [1]. First, the landmine must emit an adequate amount of the chemical signature to counter degradation and transport losses in the soil. The properties of the mine case material and method of construction are critical aspects in this most important process.

Once the landmine chemicals reach the soil, the soil acts as a storage media, releasing quantities in proportion to the sorption/desorption equilibria of the soil-water-air system. Transport or movement of the landmine chemicals occurs as a vapor in air, and as a solute in water, where diffusion and convection processes work simultaneously in a complex process. The landmine chemicals are organic molecules, which participate in soil biochemical reactions causing degradation and loss that can be very rapid under certain conditions—and very slow under others. Weather cycles dominate the driving forces that transport the landmine signature chemicals to the ground surface, where dogs identify the odors to make a positive indication.

One must recognize the dynamic variability of the landmine chemical odor, and the conditions that maximize and minimize the expression at the ground surface as a cue for the mine detection dog. Computer simulation tools have been developed that can assess these complex interdependencies and provide insight into optimal conditions for comparison to the vapor sensing thresholds of mine detection dogs.

In addition to odor sensing capability, the dog must follow instructions provided by its handler. The behavioral characteristics of an optimal mine detection dog have been identified [2], and include: nose to ground, consistency of repetitive action, obedience, endurance, focus, and slow-moving. Vapors that emanate from the ground are present in significant amounts only in a very thin air boundary layer. Beyond this layer it is believed that the vapor concentration is diluted to essentially zero. Surface soil residues of landmine chemicals are often discontinuous, which requires that the dog search consistently and slowly in a repetitive fashion over the entire area. This demands that the dog should maintain focus and have endurance to work over long field campaigns. Last, the dog must be obedient—that is, able to follow the handler's commands to start and especially to stop when dangerous conditions appear.

The sensor (the dog) response is established by traditional operant training methods from chemical cues that dog trainers provide. The source of the cue varies from chunks of military-grade TNT, to specially developed vapor sources with pure TNT, and to buried unfused landmines typical of the locale. The objective is to train the dog to lower and lower vapor sensing capabilities, consistent with what is actually found in the field. Only recently has research established the level of soil residues and vapors found in the field that will help define training aides needed to tune the dog to the greatest sensitivity possible.

Dogs indicate the presence of an odor to receive a reward, whether it be food or play. Indoctrination methods vary, reward methods vary, and results vary. This sensor (the dog's nose) and indicator system (dog training) is as varied as there are trainers. The debate among dog trainers for optimal training methods is relentless because there have been few opportunities for performance comparisons.

## CURRENT CAPABILITIES OF MINE DETECTION DOGS AND RATS

Dogs use a keen sense of smell to discriminate target odors. This sense of smell originates from ancestral survival needs to hunt for food, determine territorial boundaries, and determine friend or foe. Evolution has given us a highly developed and adaptable sensor; however, users have only begun to understand how to optimally select and field this sensor for humanitarian demining operations.

Dogs can find landmines. Field performance, however, is poorly understood. This is mostly because of the undocumented nature of conflict-based minefields. Most demining operations count the number of mines found but are unable to count the number of *missed* mines. Often, accidents from missed mines are blamed on re-mining rather than poor performance.

More recently, mine action centers are prequalifying potential mine dog organizations using test minefields. However, these have been few, the data kept proprietary and the subject of controversy. The Geneva International Centre for Humanitarian Demining (GICHD) has initiated a field research effort to define the performance of mine detection dogs, with links to environmental factors that influence the amount of landmine chemical signatures. Projects have been initiated in Sarajevo, Bosnia-Herzegovina, and in Kharga near Kabul, Afghanistan; however, these are long-term efforts that will provide the needed information only after several years of implementation.

It is unfortunate that probability of detection/false alarm rate (PD/FAR) data is not available for the dog. However, that method was developed for electromagnetic technology, where adjustments in sensing thresholds allow one to derive the familiar PD/FAR relationship. This is unlikely to be appropriate for the dog, where biological and training history contributes to variations in sensing thresholds, and environmental factors contribute to diurnal and seasonal variations in scent availability. More work is needed to define appropriate performance indicators for dogs as a group and for individuals.

The greatest advantage of the dog is in its superior ability to discriminate between different scents, making the dog an advanced multisensor. The dog is capable of detecting very low concentrations [3] and only limited work has shown types of compounds dogs use as cues [4]. But, the dog is also able to recognize multiple substances concurrently. Research has not yet defined whether one substance or a bouquet of odors is best used by the dog to discriminate the landmine from background scents. Landmines emanate a variety of substances from the main charge explosive as well as from the casing (paint, plastic, rubber, cardboard, wood, or metal). A dog or rat can be trained to detect all these odors at the same time. If there is no available odor from the main charge but there are some odor traces from the casing, a dog will detect these traces if it has been trained to do so.

## ONGOING RESEARCH EFFORTS

Rigorous research to improve deployment of mine detection dogs has been scarce. Mine dog providers and demining service providers use field operations to self-educate and improve for the next contract opportunity. Cross-organizational information sharing has been limited because of concerns over the loss of proprietary knowledge and market share. Recent cooperation sponsored by the GICHD has brought mine dog suppliers, users, mine action centers, and donors together to improve the reliability and utility of mine detection dogs. However, much more work is needed.

Currently, the GICHD study objectives are to

- develop international standards and guidelines for mine dog detection (MDD)
- facilitate/undertake targeted research to improve MDD and make it faster, safer, more reliable and predictable
- create a platform of exchange between researchers, MDD organizations, and other stakeholders.

The study has also established a global focal point for the MDD industry, one that was previously missing. Over two years, the study has evolved into many new activities, one example is the evaluation of the African Giant Pouched rats for trace chemical detection (the APOPO project). Some of the study objectives have already been addressed, including the development of international standards and guidelines for MDD and studies into breeds and tripwire detection. Other objectives in process include a comprehensive analysis of environmental effects on trace chemical detection as well as studies into training methodology, operational concepts, and Remote Explosive Scent Tracing—REST (also known as MEDDS). The latter is given high priority because of its great potential for area reduction—if proven successful.

During the past decade, much has been learned about the chemical odor from landmines, principally from a Defense Advanced Research Projects Agency program (1996–2000) that had goals to mimic the chemical sensing performance of mine detection dogs using advanced technology. This program spawned research that began to quantify the nature of the chemical signature from landmines [5,6], the rate of release of chemicals from landmines [7,8], the phase partitioning of these chemicals in soils [9], field measurements of chemical residues in soils from buried landmines [10], and simulation model estimates of vapor emanations from soils [11]. Field testing, laboratory experimentation, and simulation modeling all have shown that the chemical signature exists as an ultra-trace vapor, which challenges advanced technology applications for this very difficult problem.

## FUTURE RESEARCH NEEDS

Only recently has work been completed that has explored the chemical compounds dogs use to recognize landmines [4] and the aerodynamics of how the dog inhales vapors and aerosols [12], and compared the performance of dogs with laboratory instrumentation and detection thresholds for narcotics and other nonenergetic materials [13]. Recent evidence suggests that dogs have explosive odor thresholds a billion times less [3] than the best advanced technology the world has to offer [14].

*Fact:* Dogs find landmines, dogs miss landmines; *mystery:* What enables the dog to find landmines, what challenges dogs to find landmines? These are critical research needs for a currently functional and operational landmine detector. Past research investments in a multitude of advanced technology applications (i.e., infrared, ground-penetrating radar, electromagnetic induction) have yielded few specific improvements that have transferred to the field and increased the speed of humanitarian demining operations. Now is the time to invest in research to improve the effectiveness of the mine detection dog for humanitarian demining operations.

### Training Methodology and Operations Research

Key success factors in the use of mine detection dogs are dog/handler training methods and field operations. It is unfortunately characteristic that dogs and handlers are poorly trained—often a cause of miscommunication between the sensor and the operator. Modern technology relies on automated signal processing and alarm indications, which unfortunately are currently unsuitable for use with the dog and handler.

The training process is as varied as there are training organizations. There are people with significant training experience, but training principles tend to become corporate proprietary knowledge, leaving very little written material available to the global demining community. Lack of documentation contributes to limited institutional memory, resulting in the same mistakes being repeated. Poorly understood principles of training methodology are perhaps the greatest weakness with MDD today. The only systematic attempt to address this problem was through a study launched by the GICHD in 2000. However, more specific research is still required.

When optimal training methods are developed, one can become confident that when a sufficient landmine odor is available, the dog will indicate, the handler will observe, and the mine will be found. This still may not be adequate because field conditions affect whether landmine odors are sufficient. Field operations must take into consideration environmental factors that affect the amount of the odor present for a particular mine type (leakage rate) and diurnal/seasonal weather conditions. Selection of trace chemical detection for a particular scenario must be based in confidence that the field conditions are suitable for MDD work.

If we fully understood how dogs learn and communicate, we could reduce training time and optimize performance to make dogs more reliable detectors. We would further be able to overcome some of the problems caused by environmental factors by changing the way dogs are trained. The fact that we do not fully understand how to train and use the dogs is a great obstacle to successful use of mine dogs. More research is therefore required in the field of operational use of MDD, training methodology, and behavioral aspects for dogs and humans. Key objectives are the following:

- **Basic training methodology**—Develop training methods for dogs in specific demining tasks, such as for dogs in a REST configuration, free-running dogs with professional dog handlers, and free-running dogs to be handled by nationals in demining campaign countries.
- **Research on operational concepts**—Collect and examine data from MDD search procedures, behavior, and other elements of field operations and identify weaknesses or areas with a high potential for improvement. One such study has already been undertaken and the results revealed many surprises. One individual study is insufficient to draw conclusions and further studies should be undertaken.

## Performance Measurement and Comparison

The debate among dog trainers regarding optimal training methods has been relentless because there have been few opportunities for performance comparisons. Methods have yet to be developed to calibrate the sensitivity and substance selectivity for the dogs. It is still unknown how vapor-sensing thresholds vary between individual dogs, dog breeds, dog training programs, and for individual dogs on different days.

This further prevents adequate testing of dogs prior to field use (licensing and internal quality control), and it is an obstacle to objective and efficient investigation of missed landmine cases after MDD clearance. If there were benchmarks for acceptable detection performance, international standards could be improved to incorporate a sensitivity test prior to search. It would also be possible to determine whether the dog makes mistakes or whether lack of vapor is the reason for missed mines. Research is therefore needed to

- develop vapor-sensing performance test methods to reliably compare the result of various training methods
- measure the performance of mine detection dogs with field programs in a host country with unfused landmines and weather cycles typical of that location
- develop practical ways of measuring concentrations of target scent from spots where mines have been missed, and develop detection benchmarks for comparison.

## Remote Explosive Scent Tracing

Traditional methods with mine detection dogs are based on patterned search methods in mine suspected areas. A less common system is REST. This method relies on the capture of landmine odors on filters for later presentation to specially trained dogs. Each filter represents a sector of road or land area. This method has been successful for area reduction in a very efficient manner. One major issue, however, is that the system is poorly understood, limiting deployment to road verification.

While road verification is indeed important, the global demining process would be significantly improved if REST could be used to eliminate sectors of land (area reduction). Because area reduction is so important in humanitarian demining, REST has one of the greatest potentials for development. The system is promising, but further research is required in the following areas:

- **Training methodology**—Examine ways of training dogs to maximize detection rate and minimize false detection rate, increase search motivation and search endurance, and reduce time of training and dog/handler dependency.
- **Vapor availability**—Determine the extent of the detectable plume of scent from landmines under different circumstances (environment, landmine type, soil, and burial depth).
- **Sampling concept**—Develop a safe and reliable sampling concept where all limitations are clearly defined.
- **Filter technology**—Examine properties of filter material and optimize filters to allow highest possible interception of target scent. Further examine how to present filters to animals and vapor detectors to allow highest possible emission of scent during analysis.

## Breed Selection

Not all dogs are alike. Current use of German, Dutch, and Belgian shepherds is based on historic use of these dogs as military working dogs. Whether these breeds are indeed optimal for humanitarian demining tasks has been debated. It is further a problem that very few breeds are used, thus causing a shortage of suitable dogs for MDD. A recent study [2] concludes that there are potentially four routes to producing a mine dog. The advantages and disadvantages for each of these four routes are discussed and an attempt has been made to scale different breeds using categories relevant to the design of a mine detection dog. Eleven different breeds have been examined during this process and their strengths and weaknesses have been scaled using 14 different property indicators. The report proposes alternative breeds for use. One breed—the Swedish Drever—has been identified as particularly suitable; however, no Drevers have been trained for mine detection work to date. Research is therefore needed to experimentally train MDD dogs from alternative breeds and link deployment roles with training requirements to specific dog breeds.

## Environmental Factors

The presence or absence of the trace chemical odor from buried landmines is dependent on a complex process of release from the landmine, degradation and sorption in the soil, and volatilization from the soil surface. Many years of research have documented the fundamental properties in these processes and created simulation modeling tools [15] to evaluate the complex interdependencies among these processes. Simulation models appear accurate compared with well-controlled laboratory tests [16,17]; however, comparisons with field situations are needed. With reliable prediction of scent levels above landmines, field programs could determine whether the use of mine dogs, rats, or vapor detectors would be successful in certain areas under certain conditions. Key objectives for the research are

- **Fundamental Properties**—The initial set of data that defined the fundamental properties of trace chemical detection of buried landmines was narrow, principally to establish an initial understanding of the most sensitive processes and measurement methods. This information set needs to be expanded to include more variants for mine leakage rates, soil partitioning (soil-air and soil-water) equilibria, and biological and abiotic degradation rates specific to demining campaign locations.
- **Simulation Modeling**—Simulation modeling can provide great insight into chemical mass transport processes and the complex interdependencies in the buried landmine problem. A robust simulation modeling program is needed to define the key weather cycle, mine flux, and soil environmental conditions that define optimal and detrimental conditions for buried landmine detection. This needs to be aligned with field mine dog and trace chemical detection projects to validate the simulation modeling results.

### Landmine Detection Rats

Although dogs are known to be good scent detectors, there may be alternative animals to perform this function as well. A research project in Tanzania (APOPO) trains African Giant Pouched rats to detect landmines. Preliminary results from a comparison test between rats, REST dogs in Angola and South Africa, and free-running dogs in the United States suggest that rats are just as capable of detecting similarly low concentrations as dogs. There may, however, be many additional advantages using rats compared with dogs. Preliminary research suggests that rats are quick and easy to train and have less handler dependency than is typically found with dogs. They are small and easy to accommodate, transport, and feed. They have further proven to accommodate repetitive behavior, which typically results in better endurance and longer search. Rats are currently trained as free-running and REST rats. The latter have shown very good preliminary results, although free-running rats may also be able to compete with dogs in the future. However, much more research and practical experience is needed to determine the full potential of rats for landmine detection.

### Research Presentation (Technology Transfer)

Research alone is not enough to make changes: The results from applied research must be absorbed by those who train and use the animals. Scientific publications are likely to fail because it takes a scientist to read a scientific paper. Videos in support of scientific reports are likely to have a positive effect and enhance full understanding about MDD throughout the industry. The target group for MDD research is not necessarily the research community but the people who train and use MDD dogs. These people will better understand the optimum deployment methods, with sufficient background to make adjustments in the field, if the message is visualized through alternative information sources. Production of videos is one such source that should be further explored.

## SUMMARY

The mine detection dog is currently a valuable demining resource and has been actively in use over the past decade. Unfortunately, mine detection dogs have been fielded without significant research supporting optimum training, testing, and field conditions for deployment. The fact that the dog has succeeded in actual demining programs is a testament that the sensor is robust and is simple to use. Many of the limiting factors can be overcome with limited basic research and moderate applied research. Research investments will have tremendous impact and make significant improvements in the speed of humanitarian demining for both area reduction and individual mine detection.

## REFERENCES


1. Phelan, J. M., and S. W. Webb, *Chemical Sensing for Buried Landmines: Fundamental Processes Affecting Trace Chemical Detection,* Sandia National Laboratories, SAND2002-0909, 2002.
2. McLean, I., *Designer Dogs: Improving the Quality of Mine Detection Dogs,* Geneva, Switzerland: Geneva International Centre for Humanitarian Demining, 2001.
3. Phelan, J. M., and J. L. Barnett, "Chemical Sensing Thresholds for Mine Detection Dogs," in *Detection and Remediation Technologies for Mines and Minelike Targets VII,* J. T. Broach, R. S. Harmon, and G. J. Dobeck, eds., Seattle: International Society for Optical Engineering, 2002.
4. Johnston, J. M., M. Williams, L. P. Waggoner, C. C. Edge, R. E. Dugan, and S. F. Hallowell, "Canine Detection Odor Signatures for Mine-Related Explosives," in *Detection and Remediation Technologies for Mines and Minelike Targets III,* A. C. Dubey, J. F. Harvey, and J. Broach, eds., Seattle: International Society for Optical Engineering, 1998, pp. 490–501.
5. George, V., T. F. Jenkins, J. M. Phelan, D. C. Legget, J. Oxley, S. W. Webb, J. H. Miyares, J. H. Craig, J. Smith, and T. E. Berry, "Progress on Determining the Vapor Signature of a Buried Landmine," in *Detection and Remediation Technologies for Mines and Minelike Targets V,* A. C. Dubey, J. F. Harvey, J. T. Broach, and R. E. Dugan, eds., Seattle: International Society for Optical Engineering, 2000.

6. Jenkins, T. F., D. C. Leggett, P. H. Miyares, M. E. Walsh, T. A. Ranney, J. H. Cragin, and V. George, "Chemical Signatures of TNT-Filled Land Mines," *Talanta*, No. 54, 2001, pp. 501–513.

7. Leggett, D. C., J. H. Cragin, T. F. Jenkins, and T. A. Ranney, *Release of Explosive-Related Vapors from Landmines*, Hanover, N.H.: U.S. Army Engineer Research and Development Center—Cold Regions Research and Engineering Laboratory, ERDC-CRREL Technical Report TR-01-6, February 2001.

8. Leggett, D. C., and J. H. Cragin, *Diffusion and Flux of Explosive-Related Compounds in Plastic Mine Surrogates*, Hanover, N.H.: U.S. Army Engineer Research and Development Center—Cold Regions Research and Engineering Laboratory, ERDC-CRREL Technical Report ERDC-TR-33, in press.

9. Phelan, J. M., and J. L. Barnett, *Phase Partitioning of TNT and DNT in Soils*, Albuquerque, N.M.: Sandia National Laboratories, SAND2001-0310, February 2001.

10. Jenkins, T. F., M. E. Walsh, P. H. Miyares, J. Kopczynski, T. Ranney, V. George, J. Pennington, and T. Berry, *Analysis of Explosives-Related Chemical Signatures in Soil Samples Collected Near Buried Landmines*, U.S. Army Corps of Engineers, Engineer Research and Development Center, Report ERDC TR-00-5, August 2000.

11. Webb, S. W., and J. M. Phelan, "Effect of Diurnal and Seasonal Weather Variations on the Chemical Signatures from Buried Landmines/UXO," in *Detection and Remediation Technologies for Mines and Minelike Targets V*, A. C. Dubey, J. F. Harvey, J. Broach, and R. E. Dugan, eds., Seattle: International Society for Optical Engineering, 2000.

12. Settles, G. S., and D. A. Kester, "Aerodynamic Sampling for Landmine Trace Detection," in *Detection and Remediation Technologies for Mines and Minelike Targets VI*, A. C. Dubey, J. F. Harvey, J. T. Broach, and V. George, eds., Seattle: International Society for Optical Engineering, 2001.

13. Furton, K. G., and L. J. Myers, "The Scientific Foundation and Efficacy of the Use of Canines as Chemical Detectors for Explosives," *Talanta*, No. 54, 2001, pp. 487–500.

14. la Grone, M., M. Fisher, C. Cumming, and E. Towers, "Investigation of an Area Reduction Method for Suspected Minefields Using an Ultra-Sensitive Chemical Vapor Detector," in *Detection and Remediation Technologies for Mines and Minelike Targets VII*, J. T. Broach, R. S. Harmon, and G. J. Dobeck, eds., Seattle: International Society for Optical Engineering, 2002.

15. Webb, S. W., K. Pruess, J. M. Phelan, and S. Finsterle, "Development of a Mechanistic Model for the Movement of Chemical Signatures from Buried Landmines/UXO," in *Detection and Remediation Technologies for Mines and Minelike Targets IV,* A. C. Dubey, J. F. Harvey, J. Broach, and R. E. Dugan, eds., Seattle: International Society for Optical Engineering, 1999.

16. Phelan, J. M., M. Gozdor, S. W. Webb, and M. Cal, "Laboratory Data and Model Comparisons of the Transport of Chemical Signatures from Buried Landmines/UXO," in *Detection and Remediation Technologies for Mines and Minelike Targets V,* A. C. Dubey, J. F. Harvey, J. Broach, and R. E. Dugan, eds., Seattle: International Society for Optical Engineering, 2000.

17. Phelan, J. M., S. W. Webb, M. Gozdor, M. Cal, and J. L. Barnett, "Effect of Wetting and Drying on DNT Vapor Flux: Laboratory Data and T2TNT Model Comparisons," in *Detection and Remediation Technologies for Mines and Minelike Targets VI,* A. C. Dubey, J. F. Harvey, J. T. Broach, and R. E. Dugan, eds., Seattle: International Society for Optical Engineering, 2001.

Appendix U

# SIGNAL-PROCESSING AND SENSOR FUSION METHODS (PAPER I)

*Leslie Collins, Duke University*

This appendix focuses on the impact of signal-processing techniques on the landmine detection problem and suggests research investments that will allow continued performance improvement. The focus is primarily on processing of electromagnetic induction (EMI) data, although results for other sensors as well as sensor fusion will be discussed.

## BASIC PHYSICAL PRINCIPLES

Signal-processing algorithms for landmine detection must detect the presence of an object in the geological background and discriminate signals associated with landmines from signals associated with discrete clutter objects. In general, signal-processing algorithms perform best when the physics that define the problem are integrated within the mathematical constructs underlying the theory of signal processing and pattern recognition. Utilizing experimental data measured under realistic conditions to test the performance of algorithms, and using insight from the data to guide the algorithm development process, has proven to be crucial for reducing false alarm rates in the landmine detection problem. The utilization of computational models describing sensor phenomenology, physics-based feature selection, statistical models of mines and clutter, and spatial information have all led to dramatic reductions in the false alarm rates of landmine detection systems. Because the physics that governs each sensor modality differs, feature sets extracted from data collected by different sensors usually are not consistent across sensors. However, several common approaches to processing the raw signals or the extracted features have been applied across sensor modalities.

In the landmine detection scenario, a particular sensor, or set of sensors, is used to interrogate one or several spatial locations for which a mine/no-mine decision is to be made. The sensing process may be automated, as is the case for vehicular or autonomous systems, or may involve a human manually operating the sensor. A sensor may record all of the response defined by the phenomenology associated with that sensor, or it may only record a portion of the response. Signal-processing algorithms for landmine detection are necessarily constrained by the available sensor data. Algorithms are also constrained by the system configuration as well as their impact on operator training requirements. For example, in the Army's Handheld Standoff Mine Detection System (HSTAMIDS), a soldier operates collocated ground-penetrating radar (GPR) and metal detection sensors in two distinct modes. In a scanning mode, processing of both sensors is done via causal systems, and spatial information is not explicitly incorporated into the processing. Once one of the sensors signals the presence of a mine-like object, meaning a *detection* is made, the operator enters an investigation mode where the sensor is operated differently. In this mode, algorithms could potentially utilize spatial information and operate in a noncausal mode. The operator utilizes the information from the two sensors, as well as from visual and environmental cues, to effect *discrimination* via sensor and information fusion, whereby nuisance clutter items are potentially ignored.

## STATE OF DEVELOPMENT

In recent years, there have been substantial improvements in sensor technology, with resulting improvements in the quality of the signals available from landmine detection sensors. In the EMI regime, Johns Hopkins University has developed a high-quality time-domain system capable of recording the EMI signal very early in the response time, and Geophex Ltd. has developed a frequency-domain system that operates over a fairly broad band. Both of these sensors measure the entire sensor response and are transitioning from the laboratory to the field. In contrast, the fielded EMI sensor—the PSS-12—provides a single time sample of the EMI response curve at every spatial position sampled. Other sensor manufacturers have developed systems sensitive enough to detect the extremely low metal content of plastic mines but often do not record the entire time- or frequency-domain signature. In GPR, the recently fielded Wichmann/NIITEK radar is capable of collecting remarkably clean broadband time-domain data, and the radars that are components of HSTAMIDS (CyTerra Corporation) and the Ground Standoff Mine Detection System (GSTAMIDS) (EG&G Inc.) have demonstrated good performance in several test environments. Both seismic and quadrupole resonance sensors have also been developed and are transitioning to field tests.

In 1996, the Army Research Office funded three five-year Multidisciplinary University Research Initiatives (MURIs) to investigate phenomenological studies and signal-processing research for the humanitarian landmine detection problem. Previously, most research had been performed by government laboratories and by government contractors building systems and primarily was not basic (6.1-level) research. Prior to the MURIs, the majority of the signal processing performed in contractor systems was anomaly detection, and little if any discrimination of clutter from mines was performed. This was particularly true in EMI sensors, where energy detection was the primary mode of operation. The development of sensors that are providing better data, and the focus of the MURIs and other government-sponsored programs on advanced signal-processing and sensor fusion research, has resulted in the development and transition of algorithms that are beginning to effect discrimination, and thus positively impact the false alarm rate. The models developed under the MURIs have supported the signal-processing research. Because signal processing traditionally tends to lag sensor development, some of these algorithms are just beginning to be tested in blind tests. The MURI-based research has also resulted in an improved sense of the optimal feature set to use when processing data from the various sensor modalities, as well as performance bounds on some of the feature extraction techniques. However, there is much additional research that could be performed as additional high-quality sensor data become available, particularly in the areas of model-based signal processing and of sensor fusion. While single-sensor processing algorithms are beginning to become more sophisticated, phenomenological models have only begun to be incorporated into the processing, and most sensor fusion algorithms that have been tested in this application area are still fairly basic.

## CURRENT CAPABILITIES AND OPERATING CHARACTERISTICS

An energy detector constitutes a signal-processing algorithm that is optimal for detecting a totally random signal in a totally random background. It assumes little if any a priori knowledge of the problem but is generally robust and is thus often used as an anomaly detector. It has been shown over the last several years that advanced signal-processing algorithms can reduce the false alarm rate substantially over such simple anomaly detection strategies. These more sophisticated algorithms have benefited from phenomenological models and understanding of the signal being sensed, advances in sensor capabilities, utilization of spatial data, and selection of physically based feature sets. Because the signals sensed from mines, background, and clutter are not deterministic quantities, statistical treatment of the various signatures has also had a positive impact on discrimination performance.

Metal detectors, for example, have advanced to the point where they can detect nearly all of the metal present in the environment down to tactical landmine burial depths. However, discriminating metal in a landmine from metallic clutter is a substantially more difficult problem. Similarly, discriminating a rock from a landmine in GPR data is more difficult than discriminating a landmine under the ground from the ground itself. Techniques that have been investigated to effect discrimination include Bayesian strategies, clustering techniques, hidden Markov models, inversion, support vector machines, and fuzzy processing. In a study performed at Duke University with the GEM-3 sensor, the false alarm rate was reduced by a factor of 10 when a statistical decision strategy was used in place of an anomaly detection (energy) strategy in a blind field trial. It is also possible to improve performance with systems that record only a portion of the received signal. For example, researchers at Auburn University demonstrated that the false alarm rate associated with the PSS-12 was reduced by a factor of 4 by processing the spatial pattern associated with the received signal measured over a suspect object. Various algorithms for EMI are currently being transitioned to fielded sensors to be tested, and algorithms are also being tested with some of the newer EMI technologies.

Algorithms for some of the newer technologies, such as acoustic and quadrupole resonance sensors, are less mature than those that have been developed for EMI and radar modalities. For example, the Quantum Magnetics quadrupole resonance sensor mitigates radio frequency interference in the demodulated sensor data via a least-mean-squares algorithm and then performs the detection using an energy detector in a band of frequencies around DC. Initial research has indicated that more advanced signal-processing techniques, such as Bayesian techniques or algorithms based on spectral estimation, may provide substantial reductions in the false alarm rate. However, such techniques must be tested on larger data sets to evaluate their robustness before definitive performance comparisons can be made.

Sensor fusion in systems currently being tested by the government is still in its research infancy. HSTAMIDS uses the operator to perform sensor fusion, although joint research between the University of Missouri–Rolla, University of Florida, Duke University, and CyTerra Corporation indicated that a fairly simple processing algorithm could meet or exceed the performance of the human operator *without* using any of the spatial information assumed to be used by the operator. Among other approaches, a voting scheme is being considered for GSTAMIDS, although more sophisticated techniques are being investigated. One reason for the lack of more sophisticated sensor fusion techniques is that collocated multisensor data have only recently become available in the community.

## LIMITATIONS AND RESTRICTIONS ON CAPABILITIES

Many of the discrimination techniques, as opposed to simple anomaly detection techniques, require training data. With the advent of accurate phenomenological models, training data that accurately mimic received sensor signals from mines under a variety of environmental and soil conditions are becoming available. Such training data will aid in the analysis of the robustness of discrimination algorithms. In addition to requiring training data for mines, most algorithms will require samples of background data local to the site under test to develop the statistics or features associated with the null hypothesis. Some mechanism for incorporating a priori knowledge of the class of targets likely to be present, environmental conditions, and other site-specific parameters into the processing algorithm by the sensor operators will also be necessary.

For most sensor modalities, there will always be discrete clutter items whose signature is similar enough to the signature of a mine that they will cause false alarms. Improved sensors should aid in this problem to some degree because more information can be extracted from the sensor and utilized in the signal-processing algorithm. Sensor fusion algorithm research, in concert with continued sensor development in alternative modalities (such as quadrupole resonance), should also help address this limitation.

Discrimination algorithms also usually assume isolated anomalies, i.e., the signatures of individual items to be discriminated do not overlap. Although some preliminary studies at Duke University have indicated that there are techniques to separate overlapping signatures, this remains a difficult research problem. For highly cluttered sites, the development of robust algorithms to detect the presence of overlapping signatures and then separate the signatures prior to applying the discrimination algorithms is needed.

Other limitations include issues involving the necessity for real-time processing and training, and the requirement that algorithms are required to be simple for operators to execute. For example, Bayesian theory prescribes the optimal processor for the two-hypothesis testing problem, but this approach requires precise knowledge of the probability density functions describing the data under each hypothesis and sometimes requires a multidimensional integration. Both of these requirements could limit the applicability of the Bayesian approach in a field-deployed system. As another example, a discrimination algorithm could perform extremely well but require carefully controlled spatial data collections, which may be difficult to train an operator to perform but which might be possible with an automated system.

## ESTIMATED POTENTIAL FOR IMPROVEMENT

Dramatic performance improvements have been demonstrated over the past few years using improved signal processing that has been based on an improved understanding of the phenomenological underpinnings of the landmine detection problem. Additional performance improvements have been demonstrated via improved sensors. Because several new sensors are entering field tests, the potential for order of magnitude improvements in the speed at which landmines can be cleared is feasible.

There are several promising GPR technologies for which algorithm development is in its infancy. The Wichmann/NIITEK sensor operated in a simplistic anomaly detection (energy detection) mode is performing comparably to other radars with more advanced processing algorithms that are meeting the government exit criteria for handheld and vehicular detection systems. Preliminary tests with more advanced algorithms indicate that an order of magnitude improvement in false alarm rates at appropriate scanning rates will be achievable with this system in the two-to-seven-year time frame.

Most fielded EMI systems have not been optimized for the landmine detection problem per se and measure only a limited portion of the available signal. Initial evaluations of more advanced EMI systems utilizing statistical signal-processing algorithms in blind tests have suggested that an order of magnitude reduction in the false alarm rate is also possible with these systems. Additional tests in traditional government test sites are ongoing and appear to support the preliminary results. Quadrupole resonance is also a promising technology for use as a confirming sensor to further reduce the false alarm rate.

Several sensor technologies are near the point that they could be used to *individually* reduce the false alarm rate by an order of magnitude, particularly in realistically cluttered test sites. The combination of these technologies with appropriate sensor fusion algorithms, and confirmatory sensors, has the potential to dramatically reduce the false alarm rate and thus provide order of magnitude increases in landmine clearance rates. To achieve this goal, investments should be made in the individual sensor technologies, signal-processing research for each of the technologies, and basic research in sensor fusion. Additional care must be taken to ensure that the system-level operation is appropriate for soldiers or indigenous populations.

## OUTLINE OF A RESEARCH AND DEVELOPMENT PROGRAM

Several well-established research results should be used to guide the design of a research and development program. These include the following:

- Incorporation of the phenomenology associated with a particular sensing modality directly into the signal-processing algorithm or into the feature selection can result in substantial performance improvement.
- Advanced signal-processing techniques can improve discrimination performance over energy-based, differential-energy based, or anomaly detection techniques.
- Multisensor systems outperform single-sensor systems, even with fairly rudimentary sensor fusion algorithms.
- Multiple sensor designs within the same sensor modality can each perform well and may generate different false alarms.
- Utilization of spatial data improves performance.
- In EMI systems, the signatures from multiple objects combine approximately linearly, and can be separated under some conditions.

These results suggest parallel development of multiple sensors in concert with signal-processing algorithms, instead of isolating the research associated with each task. It also suggests considering a system-level optimization as part of the down-select for sensors and algorithms to be incorporated into the system. Substantial reductions in the false alarm rate could be achieved by leveraging and extending previous research findings, including those listed above, and by continuing a basic research program in humanitarian demining.

## REFERENCES

Most recent field tests of systems have been reported at the UXO/ Countermine Forum and/or the International Society for Optical Engineering (SPIE) Detection and Remediation Technologies for Mines and Minelike Targets Conference. Research results from the three Humanitarian Demining MURIs and other Department of Defense (DoD)–funded projects are most commonly reported at SPIE. In addition, interim and final reports from the MURIs and other DoD-supported research programs are available from the sponsors. The Joint Unexploded Ordnance Coordination Office website (www.uxocoe.brtrc.com/lib.htm) has also archived several journal articles and technical reports.

# SIGNAL-PROCESSING AND SENSOR FUSION METHODS (PAPER II)

*Paul Gader, University of Florida*

## SUMMARY

Signal processing is a necessary, fundamental component of all detection systems and can result in orders of magnitude improvement in the probability of detection (PD) versus false alarm rate (FAR) of almost any sensor system. A recent Multidisciplinary University Research Initiative (MURI) program (now ended) specifically aimed at landmines achieved dramatic improvements by developing and applying appropriate signal-processing techniques to data acquired from advanced sensors. Several of these techniques were transferred, or are currently being transferred, to industries developing prototype hardware systems for landmine detection. Basic research and development (R&D) programs specifically focused on signal processing for humanitarian demining with strong technology transfer components will greatly enhance the likelihood of further improvements with existing and developing sensing methodologies. Focused programs of this nature do not currently exist in the United States.

## BASIC PRINCIPLES

Signal processing for landmine detection seeks to exploit discrimination information in measured signals from a variety of sensors. This information can be of many types, including frequency, shape, and size. Signal processing is used to mitigate system effects, characterize and discard sensor responses to the environment, compute discriminating features for classification of sets of measurements as either mines or nonmines, and provide feedback to operators.

Signals used in mine detection are generally multidimensional. Some example systems that are very familiar to this author are described here:

**Vehicle- or Cart-Mounted Ground-Penetrating Radar (GPR)**—These systems generally consist of arrays of transmitters and receivers and produce volumes of data. Each volume element can correspond either to spatial position on the surface and time (which roughly corresponds to depth) or to spatial position and frequency. The Vehicle-Mounted Mine Detector (VMMD) and Ground Standoff Mine Detection System (GSTAMIDS) are of this type, as are the Geo-Centers Humanitarian GPR and the NIITEK-fielded Wichmann GPR. The relative and perhaps absolute position of the measurements on these systems is usually known with a fair amount of accuracy. They provide multiple looks at most small mines.

**Handheld GPR**—These systems generally consist of small numbers of transmitters and receivers and produce sequences of radar returns. Currently, the Handheld Standoff Mine Detection System (HSTAMIDS) is the primary example of such a system. Researchers at Ohio State University and researchers in Europe, for example, are investigating other designs. No positional information is contained in the HSTAMIDS radar. These systems have repeatedly demonstrated the capability for detecting small, low-metal antipersonnel mines. Reducing the FAR is the current challenge—and that requires signal processing.

**Cart-Mounted Acoustic/Seismic Systems**—In these systems, a force is applied to the ground (of course, not with sufficient magnitude to set the mine off) and the resulting vibration of the ground is measured. Methods for measuring vibration include laser, radar, and ultrasound. The signals from these systems will also be three-dimensional, but there is no depth information here—only surface spatial position and velocity. Systems such as these have demonstrated the capability to find and characterize antipersonnel mines in both limited outdoor experiments at the Joint Unexploded Ordnance Coordination Office (JUXOCO) site and in the laboratory.

**Spectral Imaging**—Images formed in various spectral bands, usually some region in the infrared, can provide useful information, not only for mines but also for tripwires associated with mines. These systems can provide sequences of two-dimensional images. Often the images have value in several spectral bands.

## STATE OF DEVELOPMENT

Signal-processing methods are currently being developed and implemented in both academic and industrial settings and are under evaluation in both laboratory and field settings. Many techniques are developed in universities and small companies and undergo transfer in some form to private companies engaged in sensor system development.

A very productive mode of operation has been to perform basic research in the university and transfer the results to industry. The university research relies on data collected by systems under development by the government. When signal-processing methods show promise, they are presented to industry and government. The methods, often in a modified, refined form, are adopted by industries and incorporated into their systems.

A very generic view of a typical signal-processing algorithm for demining is shown in Figure V.1.

Preprocessing can be extremely important and involves tracking the stationary and nonstationary statistics of the background and subsequently removing the background and normalizing the data, decon-

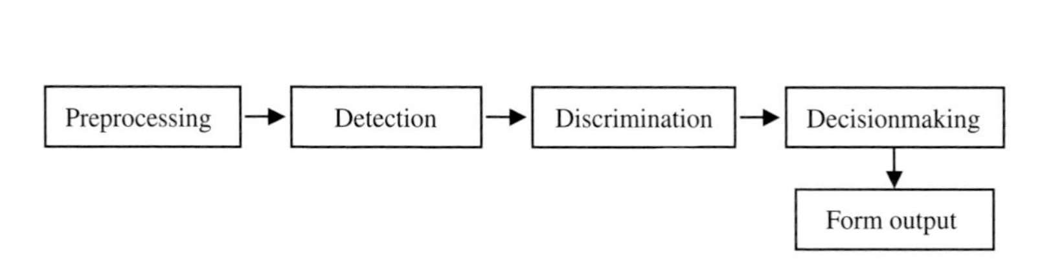


**Figure V.1—Typical Signal-Processing Algorithm for Landmine Detection**

volution of the system response to the environment, removal of interferents such as radio frequency interference (RFI), and so on. In many systems, the response of the mines cannot be seen in visualizations of the data before preprocessing but are very apparent afterward. Preprocessing is by necessity adaptive, and system performance can degrade dramatically when the adaptive estimation methods fail.

Detection algorithms generally use preprocessed data and identify regions of interest that are then sometimes used for subsequent discrimination processing. Some detection algorithms are essentially anomaly detectors that look for differences from the background. That is, once the background has been removed, any remaining signal that is "sufficiently different" from the background is considered to be a mine. Constant false alarm rate algorithms are of this type. These algorithms must be adaptive. Other algorithms, such as hidden Markov models (HMMs), attempt to model the types of anomalies that constitute mines. These algorithms can be considered both detection and discrimination algorithms.

Discrimination algorithms try to model mines, and sometimes background and clutter, to characterize the mines. HMMs, matched filters, Generalized Likelihood Ratio Tests, Linear Discriminants, Support Vector Machines, and neural networks are all examples of these types of algorithms.

Both detection and discrimination algorithms often rely on the use of features. They can be computed mathematically, as is the case in Linear Discriminant Analysis, Principal and Independent Component Analysis, and wavelets, or they may be based on qualitative/physical knowledge.

Decisionmaking involves post-processing—producing output responses for sensor fusion and for human operators. Post-processing can be used to reject responses based on such gross aggregate properties as size and shape. Responses may be produced for an operator, which is extremely important, or for the other sensors in multisensor fusion. If the system is not completely automated, i.e., responses are prepared for other algorithms or operators, then the algorithms must be more aggressive. Sensor fusion should generally require some quantitative knowledge.

## CURRENT CAPABILITIES

Dramatic improvements have been achieved using signal processing on a variety of systems. The basic university research to developmental industry research has been a productive mode, resulting in tech transfer activities at various levels of completion in the VMMD, GSTAMIDS, University of Mississippi, HSTAMIDS Countermine and Humanitarian, Geo-Centers Humanitarian, and NIITEK Wichmann systems.

Early in the previous MURI, fuzzy clustering algorithms using the Defense Advanced Research Projects Agency (DARPA) backgrounds data reduced FARs from around 30 percent to around 4 percent as shown in Figure V.2.

That algorithm was refined and implemented in real time to the Geo-Centers VMMD system, which exceeded the exit criteria at the Advanced Technology Demonstrations in 1998, achieving over 90-percent detection at approximately 0.04 FAR in field testing on antitank mines. New algorithms based on HMMs improved on that performance in the lab, achieving performance of approximately 95-percent PD at approximately 0.02 FAR in the lab, again on antitank mines. This algorithm is currently undergoing tech transfer to the GSTAMIDS system and is currently achieving similar scores of

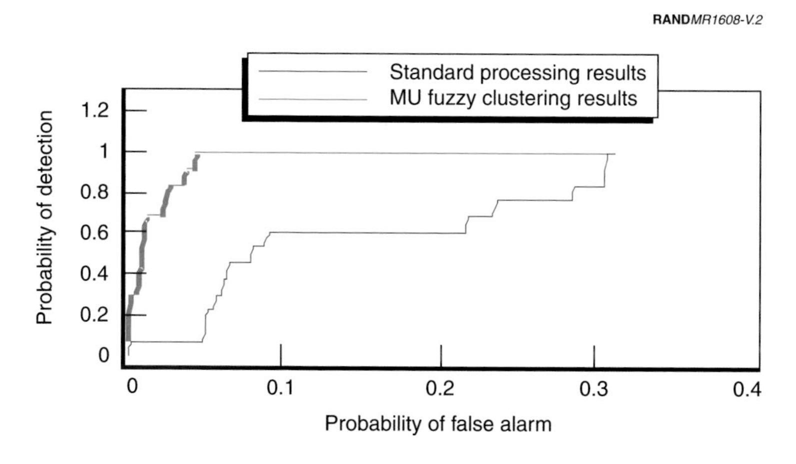


**Figure V.2—Results of Signal Processing on DARPA Background Data**

around 95-percent PD at about 0.02 FAR in the laboratory. The HMM algorithms are currently being modified and implemented on the Geo-Centers Humanitarian system. HMMs show great promise, not only in this application but in several other applications in landmine detection.

Handheld systems, such as HSTAMIDS, are very important for the humanitarian demining mission. These systems have been tested, and the results are available in government reports. In addition, preliminary work with spatial algorithms has been implemented that decreases the FAR from 38 percent to 9 percent on the calibration area of the JUXOCO grid (see Figure V.3). When combined with an electromagnetic induction (EMI) processing algorithm, the performance of the automated spatial algorithm exceeded that of a human operator (see Appendix U). Thus, spatial signal processing has definite potential for reducing FARs and possibly reducing reliance on the expertise of the operator.

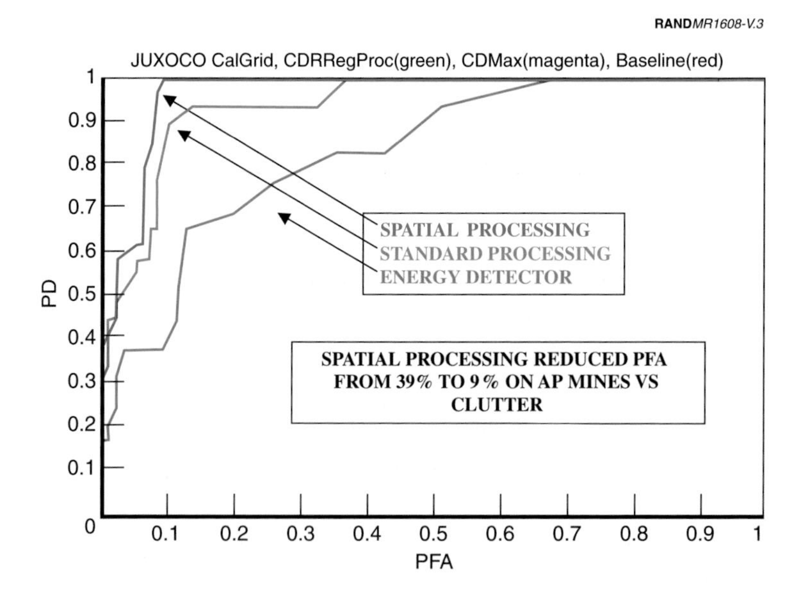


**Figure V.3—Reduction in FAR Obtained from Spatial Processing of GPR Data Collected from a Handheld Unit**

Image-processing algorithms have been developed for the acoustic-seismic system developed by the University of Mississippi. Data were acquired from some squares of the JUXOCO calibration grid and over grid squares containing low-metal antipersonnel mines and clutter objects, such as wood and plastic. The problem in this case was discriminating the antipersonnel mines from the clutter objects. Using a leave-one-out testing method, the algorithm outperformed the human experts, achieving the results shown in Figure V.4.

Signal-processing algorithms have consistently resulted in very significant drops in false alarm rates while maintaining probabilities of detection and can outperform expert human operators. Based on the experience of the past several years, this trend has not let up, and we expect these kinds of improvements to continue as new types of sensors and more advanced traditional sensors are developed.

RANDMR1608-V.4

(a) Using all

| | FA | Mine |
|---|---|---|
| FA | 27 | 0 |
| Mine | 1 | 11 |

(b) Leave-one-out

| | FA | Mine |
|---|---|---|
| FA | 27 | 0 |
| Mine | 1 | 11 |

(c) Human expert

| | FA | Mine |
|---|---|---|
| FA | 23 | 4 |
| Mine | 3 | 9 |

**Figure V.4—Classification Results on the Acoustics Data**

## LIMITATIONS

Further R&D is required to achieve desired performance levels. Some limitations are discussed here. Although the limitations are grouped into discrete categories, they do overlap.

### Limited Training and Testing Data

Efficient gathering, storing, and indexing of data sets is crucial for signal-processing algorithm development and analysis for humanitarian demining. Orders of magnitude improvements have been achieved through statistical and other techniques that require sufficient training data. The size of training, validation, and testing data sets may be quite large. A typical two-to-three-day data collection may result in many gigabytes of data. The ground truth for these data is often only partially accurate, requiring significant effort in ascertaining the exact location of the mine signatures in the data. Many such data sets are required for robust algorithm development.

A few standards are in place, such as the Countermine Test Management System truth file format, but no standard exists for linking the truth file format to data files produced by sensor data collection systems. There have also been some attempts at creating data repositories, but they are limited.

### Discrimination in High Dimensions

Many of the dramatic advances in the past few years have come from better anomaly detection. Many simple algorithms detect any anomalous behavior in measured data. Much of the anomalous behavior is due to system cross-talk and nonstationary effects of the ambient environment, such as the ground and temperature of the equipment. The result was that many algorithms declared mines present when there was no object present. Detection algorithms have become much better at eliminating these kinds of problems. In addition, new sensor systems, such as the Wichmann GPR, have much cleaner signals, reducing the probability of these anomalies. We are currently at the point that region-of-interest detectors can be used to discard most of the measurements that are not associated with mines. The remaining measurements must be processed more carefully to discriminate mines and nonmines.

Thus, the problem in some cases has shifted more to one of discriminating between objects and is similar in nature to pattern recognition problems. A standard pattern recognition problem is to assume that a set of features has been measured from a region of interest and that one seeks to categorize the features as belonging to a finite set of classes.

In the case of landmines, we wish to measure features on sensor data and categorize the features as mine or not mine. The distinction here is that the set of objects that are not mines is not a coherent class. Thus, one cannot estimate the distribution of these objects or really define them as a class. (This problem is not isolated to landmines but occurs in many other real-world pattern recognition problems. It is not a solved problem.)

This implies the need to characterize precisely the class of mines. This is a more difficult task than it might seem because most representations of mines via sensor data require high-dimensional representations. Mine signatures can be highly variable, so complicated regions in feature space are required. We often use low-dimensional intuition to guide the development of high-dimensional methods. However, regions in high dimensions can be counterintuitive. For example, in a unit cube in high dimensions, the distance to the side is constant as a function of dimension, but the distance to the corner goes to infinity as the dimension goes to infinity. Another example of counterintuitive behavior is given by the Busemann-Petty conjecture. It seems reasonable to assume that if every symmetric, convex central slice through object A is bigger than every symmetric, convex central slice through object B, then object A is bigger than object B. However, this is only true for objects with dimensions less than 5. The characterization of class regions in high-dimensional space is still not well understood, but it is important for discriminating mines from clutter.

As another example, it is known that multilayer perceptions (MLPs) are excellent discriminants if the input pattern is known to come from a fixed, finite set of classes. However, an MLP cannot create a closed region in feature space unless the number of hidden nodes is greater than the number of input features. Even if this property holds, it is not guaranteed to form a closed region, and it is an extremely hard problem to check if the region is closed. Thus, MLPs are probably not reasonable for landmine discrimination problems.

Some methods, such as robust or possibilistic clustering, self-organizing feature maps, and relevance vector machines, provide promise of helping to achieve better understanding and high-dimensional modeling and hold promise for improved discrimination capabilities for many sensors.

## Integration of Physical Models and Algorithms

Physical models offer great promise in helping to identify discriminating features, to identify bounds on variability of signatures, and even to create synthetic training sets. The results of physical models can be qualitative understanding of the phenomenology associated with a sensor, parametric mathematical models that can model the range of responses from a sensor, or actually synthetic "raw" data. Some examples of qualitative features that are currently under investigation are the "double humped" form of the returns in forward-looking radar and the transmission zeros found in acoustic-seismic systems. The prototypical mathematical models are the decay curves for EMI. Some synthetic data have been generated for multifrequency EMI and the Wichmann GPR, but conclusive results on robust accuracy are not known by this author. These methods are very computationally intensive. Continued research in this area is important.

## System-Level Optimization

Detection algorithms consist of a sequence of steps. Often each step is developed independently and optimized relative to short-term criteria. For example, preprocessing algorithms are generally evaluated based on reduction of "noise," which is sometimes quantitative and sometimes qualitative but almost never related to the end goal of PD versus FAR. We have found that in aided target recognition and other pattern analysis problems, "noise" at times can be anomalies associated with objects of interest and the process of cleaning the noise so that the data "look" better can reduce performance. Algorithms that consist of multiple steps should be optimized as a system, using the end goal as the objective. Unfortunately, these objectives and algorithms are not well behaved in a mathematical sense. The algorithms are complex, highly nonlinear sequences of steps. The relationships of the input distributions to the output distributions are generally extremely difficult or impossible to compute. Gradient descent on the objective function can also be difficult because gradients tend to vanish in multistage, complex systems. Stochastic search algorithms offer promise in this area. System-level optimization is an important but difficult unsolved problem in the development of signal-processing algorithms for humanitarian demining.

## Adaptive Processing

Adaptive processing is crucial to landmine detection. Humanitarian systems may need to operate for many hours at a time over a range of ground and atmospheric conditions. Data acquired by most sensors are highly nonstationary. The data may change rapidly or slowly, and the density of mines may be high or low. Adaptive methods have been used in all signal-processing algorithms that we know of that have been transferred from basic research to industry. These methods generally adapt the statistics of the background. Currently, if the mine density is high, they will fail; they can also fail when conditions change too rapidly. These methods have not been tested over long periods.

In addition to background removal, adaptive processing is required for removal of interferents, such as RFI for quadrupole resonance detection. This is a very important area of research.

Multisensor fusion should be adaptive. Different sensors are useful in different conditions, and a dramatic change in behavior of one sensor system may not be present in another sensor system, providing a sort of "check and balance" system. Continued research is needed that develops more sophisticated adaptive methods for demining.

## Incorporation of Spatial Information in Handheld Systems

Handheld systems are very relevant to the humanitarian problem. Mines have spatial signatures; they are not one-dimensional. Currently, there are few algorithms for incorporating spatial information into the algorithms on these systems. Incorporation of spatial information is difficult because the sampling in handheld systems is time-based and there is no machine control of the position of the sensor when samples are collected. That is, samples are collected at fixed time intervals over an irregular set of points on the surface. Preliminary work with HSTAMIDS data demonstrates that spatial information can dramatically improve performance.

## Incorporation of User Feedback

Currently, very few systems incorporate user feedback. HSTAMIDS allows the user to turn off adaptation when a potential mine has been encountered. This is a simple form of user feedback. Certainly, user feedback must be simple enough to provide for use by a wide variety of operators, but this may limit the types of feedback that can be used. Consider a case in which a deminer has just found a mine. The data for that mine could be used to update the signal-processing algorithm to improve the capabilities of the system in order to find mines in the current environment.

# POTENTIAL FOR IMPROVEMENT

Signal processing is a necessary and fundamental component of all detection systems and can result in orders of magnitude improvement in the PD versus FAR of almost any sensor system. Continued focused research in this domain is necessary to reach higher levels of performance.

# OVERCOMING CURRENT LIMITATIONS: SENSIBLE R&D PROGRAM

A basic research program specifically aimed at signal processing for landmine detection should be created. Excellent results were achieved in the past, but there is currently no focused basic research program.

Primary goals of R&D in signal processing should be the development of new techniques, education, and technology transfer. Subsidiary goals should be the establishment of demining specific, user-friendly software tools in a MATLAB/C environment and standardized databases useful for entire communities of researchers. These goals are not mutually disjoint. The content of the research should address the limitations mentioned above.

Development of new techniques involves not only new signal-processing algorithms but also new methodologies for applying signal-processing concepts to the humanitarian problem. The academic literature is rich with ideas but short on detailed examination of the applicability of these ideas to large-scale, real-world problems, such as humanitarian demining. Such basic research is best carried out by combining academic research with industrial R&D.

Education involves graduate-level courses at universities, short courses at industrial and government locations and conferences, small workshops involving multidisciplinary teams, and production of written materials that are tutorial or educational in nature. Education also overlaps significantly with technology transfer.

Technology transfer often begins with university or small R&D company researchers using data from prototype systems developed by industrial contractors. Techniques are developed and demonstrated in the lab using these data. They are then presented to the government and the industrial contractors. If techniques display success in the lab, then computer programs and descriptions can be provided to the industrial contractors as well as assistance in interpreting and implementing them within the real-time system environment. Industry developers often modify and refine techniques as a result of further evaluation and testing.

Research at universities and small R&D firms should be supported directly by the government to ensure broad applicability of new methodologies. Incentives for industry to work with basic researchers are also necessary.

## BIBLIOGRAPHY

Gader, P. D., J. M. Keller, and B. N. Nelson, "Recognition Technology for the Detection of Buried Land Mines," *IEEE Transactions on Fuzzy Systems*, Vol. 9, No. 1, February 2001, pp. 31–43.

Gader, P. D., M. Mystkowski, and Y. Zhao, "Landmine Detection with Ground Penetrating Radar Using Hidden Markov Models," *IEEE Transactions on Geoscience and Remote Sensing*, Vol. 39, No. 6, June 2001, pp. 1231–1244.

Gader, P. D., B. Nelson, H. Frigui, G. Vaillette, and J. Keller, "Landmine Detection in Ground Penetrating Radar Using Fuzzy Logic," *Signal Processing, Special Issue on Fuzzy Logic in Signal Processing*, Vol. 80, No. 6, June 2000, pp. 1069–1084.

Gori, M., and F. Scarselli, "Are Multiplayer Perceptrons Adequate for Pattern Recognition and Verification?" *IEEE Transactions on Pattern Analysis and Machine Intelligence*, Vol. 20, No. 11, November 1998, pp. 1121–1132.

Ho, K. C., and P. D. Gader, "A Linear Prediction Land Mine Detection Algorithm for Hand Held Ground Penetrating Radar," *IEEE Transactions on Geoscience and Remote Sensing*, Vol. 40, No. 6, June 2002, pp. 1374–1385.

Ho, K. C., P. D. Gader, and J. B. Devaney, "Locate Mode Processing for Hand-Held Landmine Detection Using GPR," in *Detection and Remediation Technologies for Mines and Minelike Targets VII*, J. T. Broach, R. S. Harmon, and G. J. Dobeck, eds., Seattle: International Society for Optical Engineering, April 2002.

Hocaoglu, A. K., P. D. Gader, J. M. Keller, and B. N. Nelson, "Anti-Personnel Land Mine Detection and Discrimination Using Acoustic Data," *Journal of Subsurface Sensing Technologies and Applications*, Vol. 3, No. 2, April 2002, pp. 75–93.

Keller, J. M., P. D. Gader, Z. Cheng, and A. K. Hocaoglu, "Fourier Descriptor Features for Acoustic Landmine Detection," in *Detection and Remediation Technologies for Mines and Minelike Targets VII*, J. T. Broach, R. S. Harmon, and G. J. Dobeck, eds., Seattle: International Society for Optical Engineering, April 2002.

Rotundo, F., T. Altshuler, E. Rosen, C. Dion-Schwarz, and E. Ayers, *Report on the Advanced Technology Demonstration of the Vehicular Mounted Mine Detection Systems,* IDA Document D-2203, October 1998.

Stanley, R. J., S. Somanchi, and P. D. Gader, "The Impact of Weighted Density Distribution Function Features on Landmine Detection Using Hand-Held Units," in *Detection and Remediation Technologies for Mines and Minelike Targets VII*, J. T. Broach, R. S. Harmon, and G. J. Dobeck, eds., Seattle: International Society for Optical Engineering, April 2002.

Appendix W

# CONTACT METHODS

*Kevin Russell, Defence R&D Canada–Suffield*[1]

## INTRODUCTION

Although a small number of landmines were introduced into modern warfare during World War I, the tactics of landmine use, both in the deployment and removal, did not become clear until World War II. By 1939, the German and Italian armies had developed both antitank and antipersonnel landmines, which were used effectively against the Allied forces. Naturally, the Allies developed techniques for defeating the defensive barricade presented by a well-laid-out minefield. The obvious array of flails, rollers, and projected charges were used. However, "the first, and throughout the entire war, the commonest procedure was to locate the mine, neutralize it if necessary and remove it by hand"[1]. Although metal detectors based on the principle of heterodyne oscillation were used to rapidly scan an area, the final approach to a landmine prior to neutralization inevitably used a pointed stick, such as a bayonet.

As discussed in Russell [1], "a bayonet, held obliquely in the hand and prodded into the ground in an arc-like pattern, was the best available means of mine location. Many a sapper or infantryman played out his luck when the bayonet struck the prongs of an S-mine." The S-mine 35 was a bounding cylindrical steel German antipersonnel mine, four inches in diameter and five inches high. At the time, metal cased landmines were not the only threat. The Italian antipersonnel landmine was a small Bakelite box, and the Russian

[1]Originally published by Defence R&D Canada–Suffield. Reprinted with permission.

army fielded several landmines of low metallic content. In the final years of World War II, various metal detectors were developed using the principles of heterodyne oscillation, super-regeneration, and the well-known inductive bridge (first used in World War I). Since then, the conventional metal detector has seen some technological improvement that has increased the sensitivity and improved the ergonomics of fielded systems.

Conversely, the tool used to precisely localize a landmine remains a derivative of the pointed stick. Very little advancement has been made in regard to the soldier's bayonet. The modern military uses a conventional lightweight nonmagnetic probe (prodder). Humanitarian demining organizations use a wide variety of tools that range from the military prodder to the ordinary screwdriver or some locally fabricated device consisting of a metal rod and a wooden handle. Although the tool may vary, the technique remains the same. Once a suspect area has been localized with a metal detector or some other similar tool, a person with a prodder gingerly probes and excavates the ground until a positive and unequivocal identification can be made of the buried object.

It is still a researcher's dream to wave a Star Trek tricorder at a heavily forested area and produce a detailed map of all the unseen hazards. If, in fact, this were possible, the final task would still remain; the buried objects would require precise localization prior to removal or neutralization. Although one can envision the use of mechanical equipment to remove intact, undetonated landmines, the logistical support is usually problematic in third-world countries. As a result, mechanized removal systems have not been fielded by demining organizations.

## BASIC PRINCIPLES

Manual prodding and excavating is conceptually a simple process. Simple tools, such as a trowel and a screwdriver, are used to gently probe the ground until a solid object is contacted. Material surrounding the buried object is carefully removed until a positive identification can be made. This process is repeated for each buried object until the area is cleared.

As described in Gasser and Thomas [2] and Gasser's *Technology for Humanitarian Landmine Clearance* [3], the human operator is intimately involved in the prodding process. Landmines are typically activated by pressure fuses and this requires the operator to limit the amount of force applied to the prodding tool. In many cases, the operator exceeds the force required to activate the landmine [4], but fortunately the contact point of the prodding tool is not usually on the fuse mechanism.

## STATE OF DEVELOPMENT

The conventional prodder has been improved from the original "soldier's bayonet" to a lightweight, nonmagnetic, and wear-resistant instrument. There are many styles of these prodders available from such companies as RUAG Munition (Switzerland), Ribbands Explosive (United Kingdom), and Dyno Nobel (Denmark) with minor variation in the handle, length, and hand protection. Some, such as the HARC #3 from the University of Western Australia, are a combination of a prodding and digging tool with limited blast protection.

There have been some attempts to increase the sophistication of the conventional prodder. For example, the Croatian army introduced a hollow tube prodder where the impact noise of the probe tip contacting a buried object could be easily heard. In addition, the operator could scrape the surface of the buried object. Different material would emit characteristic sounds and the operator could use this information to make an educated guess about the buried object.

More advanced prodders have been developed. The use of the "feedback prodder" in Afghanistan was reported by Gasser [4]. He discovered that deminers "(1) repeatedly used more force than is required to activate some mines and (2) consistently underestimated the force they were using by large amounts, often thinking they were using about half the actual force." Gasser concluded that an improved feedback prodder could be a valuable training tool for the development of prodding techniques that limit the force applied by the operator.

For a brief period, DEW Engineering and Development Ltd. (www.dew.ca) manufactured a prodder with an ability to discriminate between plastic, rock, and metal. The SmartProbe™ was based on technology developed at Defence R&D Canada (www.suffield.drdc-rddc.gc.ca) where an acoustic pulse is used to characterize the material under contact.[2] The device was tested at the Cambodia Mine Action Center in 1999 with positive feedback from the deminers. The U.S. Department of Defense (DoD) Humanitarian Demining Technologies Program also evaluated the SmartProbe with mixed results [5]. The Canadian Armed Forces conducted the last known test of DEW's prodder in September 1999 [6]. Although the Canadian army highly rated the concept of operations, several shortcomings in the ergonomic design, ruggedness, and performance discouraged the acquisition of the SmartProbe for field use.

The Canadian Centre for Mine Action Technologies (CCMAT) supported HF Research Inc. (www.hfresearch.com) in 2001 to improve upon DEW's SmartProbe. HF Research combined the acoustic pulse with a force feedback system in an attempt to address some of the performance limitations. The prodder was tested at CCMAT's facilities in September 2001 with very promising results [7].

## CURRENT CAPABILITIES AND OPERATING CHARACTERISTICS

The current capability of the conventional prodder is excellent. A well-defined user community has directed the design from the original "bayonet" to the modern lightweight, nonmagnetic, and wear-resistant prodding and digging tool. Useful prodders can also be fabricated in-country by local deminers using readily available materials.

As measured in Melville [6], the conventional prodder detected 100 percent of the buried objects at a rate of 1.58 sq m per hour. The ground was easy to prod and the soldiers used the standard Canadian Forces prodding technique (2-cm prodding grid and 30° prodding angle in combat dress). The probability of detection (PD) and the false alarm rate (FAR) are meaningless when evaluating the

---

[2]U.S. patents 5754494, 5920520, 6023976, and 6109112; Canadian patents 2218461 and 2273225.

conventional prodder because they merely represent the distribution of targets versus nontarget objects. However, that being said, the PD, FAR, and the probability of false alarm (PFA) can be calculated and the results are presented here. All 38 landmines were found (100-percent PD) as compared with the 119 rocks (FAR of 4.1 FA per square meter) that were placed in the lanes. The soldiers were also given the opportunity to identify the buried object before it was uncovered. They correctly declared 20 out of 38 objects as mines and 110 out of 119 objects as rocks (PD ≈ 53 percent and PFA ≈ 8 percent).

Advanced or instrumented prodders are still in their infancy. As discussed in *Technology for Humanitarian Landmine Clearance* [3], instrumented prodders provide an opportunity for very "close-in" location and discrimination. The prodder is capable of delivering sensors to close proximity of buried landmines. Instruments based on the principles of acoustics, electromagnetics, thermal conductivities, chemical analysis, and nuclear techniques can all benefit by the reduced standoff offered by the prodder. For example, chemical analysis techniques using a prodder as a sampling tool would greatly benefit from the relative abundance of explosives in the soil as compared with explosive vapors found at the soil surface. In addition, other technologies, such as electrical conductivity and spectral analysis, can be used when the probe is in contact with an object.

The DEW SmartProbe demonstrated in Melville [6] a discrimination capability where 84 of 105 objects were correctly identified as mines (PD ≈ 80 percent). Rocks were correctly identified 298 times out of 468 encounters (PFA ≈ 36 percent). The SmartProbe had an advance rate of 1.31 sq m per hour. The Canadian Forces test results agreed with laboratory measurements provided by the manufacturer. Inexplicably, the tests conducted by the DoD Humanitarian Demining Technologies Program produced drastically different results. One hundred forty two of 205 objects were correctly declared mines (PD ≈ 69 percent) and 2 of 50 objects were correctly declared rocks (PFA ≈ 96 percent).

HF Research's improvements were tested using a bench prototype prodder in September 2001 [7]. The improved prodder correctly declared 256 of 264 objects as mines (PD ≈ 97 percent) and identified 480 objects out of 792 as rocks (PFA ≈ 39 percent). The report indicates that the eight missed mines were large rusted metal mine surrogates. None of the plastic or wooden mines was missed. The advance rate of the improved prodder was not measured. However, because the operator is not required to change his or her grip to see the indicator LED [light-emitting diode] as on the DEW SmartProbe, it is expected that the improved prodder has a faster advance rate.

The prodder advance rate suggested by Melville [6] for both the conventional prodder and the SmartProbe should only be used as a preliminary indicator. The soldiers involved in the tests did not treat the unknown objects as hazardous after a declaration was made, and the object was quickly uncovered, identified, and recorded. In any case, the advance rate was dominated by the 2-cm prodding grid (≈ 2,500 prods per square meter).

## KNOWN OR SUSPECTED LIMITATIONS OR RESTRICTIONS

The conventional prodder is limited by the available training tools and the lack of a discrimination or precise localization capability. In addition, manual prodding is limited by certain environmental factors, such as ground hardness and dense root structures. A few manufacturers do consider the possible side effects of the prodder becoming a deadly projectile upon any accidental detonation of the landmine. However, a large percentage of the available prodders do not gracefully react to a mine blast and result in more severe injuries to the deminer.

Instrumented prodders are limited by the imagination of the academic and demining community. It is technically possible to construct a prodder with an electromagnetic induction system that will allow the precise localization of small metal fragments. The design of conventional metal detectors allows the operator to localize a small signal to an area under the detector head, usually about 100 sq cm. However, a significant amount of time is spent prodding the area to locate the source of the signal. In some occasions, the soil is sifted through the metal detector head to find the small metal fragment.

The instrumented prodders from DEW and HF Research use the acoustic impedance mismatch between differing materials as the basis for their discriminating capability. The mismatch is directly related to the material hardness and the contact pressure. Plastic and wood are softer than rocks, while metals are generally harder than rocks. The classification boundary between rocks and light metals (such as aluminum) is not well separated and this is one of the limiting factors of the acoustic prodder.

One of the largest barriers to the adoption of instrumented prodders by the demining community is the rigid adherence to existing operating procedures. The demining community views the current operating procedures as "safe" and resists the introduction of new equipment that does not have a preexisting safety record. In addition, although a demining organization is technically tasked to remove all the buried ordnance, quality control is usually based on random spot checking with a handheld metal detector. As a result, some demining organizations concentrate on removing all pieces of metal. It is unclear if an instrumented prodder based on an acoustic principle would assist in this task.

## ESTIMATED POTENTIAL (TWO TO SEVEN YEARS)

Other than the "feedback prodder" proposed by Gasser [4], training of the operator is provided through the observations of an instructor. As suggested by Gasser and observed by HF Research [7], the operator of a prodder is a critical part of the system.

> Manual prodding/excavating is commonly used to find and identify the mines/[unexploded ordnance] and metal fragments initially located by metal detectors, although it is frequently regarded as outmoded, unsophisticated, dangerous and in urgent need of replacement. A more careful investigation reveals that it is a very subtle and complex process, and humans are extremely well adapted to performing this task which involves fine tactile control with simultaneous observation and decision making. [3]

Improvements to the training tools, such as force indicators or pressure-sensitive dummy mines, can be realized within a few years.

The current standards used to develop NATO and other armed forces' prodders can be expanded to include a testing methodology that considers the effects of an accidental blast. Current activity within the International Test and Evaluation Program has demonstrated that an antipersonnel mine strike against a deminer during the prodding task is survivable with minor injuries, given a small blast mine and a prodder that deforms gracefully under the force of the blast. The results of these activities can be integrated into existing International Mine Action Standards.

A survey of the demining community can guide the development of instrumented prodders. The survey can be used to guide the creation of an acceptable operating procedure where an instrumented prodder can be utilized to its full potential. Although not all demining situations can benefit from an improved prodder, it will be useful in some cases. It is important to know what type of instrumented prodder could be used in which situation. A comprehensive user survey of existing demining organizations can be completed within a year.

Basic research in other technologies for very "close-in" technologies can realize significant gains over the next few years. Traditional military research programs tend to concentrate on various orders of standoff capabilities with the goal of eliminating the landmine threat while traveling at high speeds (greater than 20 km per hour). Active deminers within humanitarian organizations routinely work very carefully and slowly within inches of the explosives. Improvements to "prodder-like" tools can immediately aid the deminer by providing improved safety, localization, and discrimination. For example, the existing instrumented prodder could be augmented with a combination of a ground-penetrating radar, an electrical conductivity sensor, and/or an electromagnetic induction system to aid in the localization process. In addition, the prodder is an excellent tool to deliver confirmation sensors, which have the ability to detect explosives (chemical analysis, various nuclear interaction techniques, and nuclear quadrupole resonance).

## OUTLINE OF A RESEARCH AND DEVELOPMENT PROGRAM

While it is possible to propose some research and development (R&D) programs without considering details, such as the end user, it is not possible in the case of the prodder. The interaction between the prodder and the deminer is extremely complex and, as a result, the R&D program cannot work without user input. As such, the following outline does not propose basic theoretical research but concentrates on delivering existing technology and know-how into the hands of deminers. Although difficult to estimate, previous experience has shown the following tasks could be completed within seven years for approximately $6 million. It should be noted that engineering development is a significant portion of the estimate. Manufacturers are reluctant to invest heavily in equipment development because of the limited market potential provided by the demining community. Instead, manufacturers rely on donor organizations' or governments' support to develop equipment that is not destined for the military market.

1. Investigate the possibility of a standard for testing landmine prodders. The mechanical construction, localization, and discrimination capability should be considered by the standard. This will provide a benchmark by which the deminers can evaluate the prodder before making procurements and attempting to use the equipment in the field. It will also guide the manufacturers by providing an acceptable baseline for their development.

2. Survey the user community to develop effective operating procedures for instrumented prodders. In the foreseeable future, the deminer will remain "in the loop" while using some form of scanning/confirmation detector along with a probing/digging tool. A successful prodder cannot be developed without considering the system within which it operates.

3. Design a training package for the conventional prodder. The training package should contain tools that allow the instructor to measure the effectiveness of the students. Force indicating prodders or pressure-sensitive dummy mines are possible items for the training kit.

4. Based on the survey of the user community, develop a ruggedized version of HF Research's instrumented prodder. As discussed in the company's report [7], several technical challenges remain.

5. Investigate the possibility of incorporating a ground-penetrating radar, an electrical conductivity sensor and/or an electromagnetic induction system into a prodder-like package. Each of these technologies, including the existing instrumented prodder, are well understood, but the major difficulty involves the development of the appropriate ruggedized package.

## REFERENCES


1. L. Russell, "Detection of Land Mines (Part I)," *Canadian Army Journal*, No. 1, November 1947. Available on a CD-ROM entitled "Information Warehouse (LLIW/DDLR) Version 4.0," distributed by the Army Lessons Learned Centre of the Department of National Defence of Canada.

2. R. Gasser and T. Thomas, "Prodding to Detect Mines: A Technique with a Future," in *2nd IEE International Conference on Detection of Abandoned Landmines*, Edinburgh, United Kingdom, October 1998.

3. R. Gasser, *Technology for Humanitarian Landmine Clearance*, Ph.D. thesis, University of Warwick, Coventry, United Kingdom, 2000.

4. R. Gasser, "Feedback Prodders: A Training Tool to Improve Deminer Safety," *Journal of Mine Action*, No. 5.2, 2001. Available at http://maic.jmu.edu/journal/5.2/features/prodders.html.

5. U.S. Department of Defense Humanitarian Demining Technologies Program, *Operational Capabilities Demonstration and Test Report: Smart Mine Probe*, U.S. Army Communications and Electronics Command, Night Vision and Electronic Sensors Directorate, Fort Belvoir, Va., August 1999.

6. D. Melville, *User Evaluation Report: SmartProbe Instrumented Prodder*, Land Forces Trials and Evaluation Unit, Oromocto, Canada, October 1999.

7. HF Research Inc., *Force and Temperature Compensation for the Instrumented Prodder*, Contract Report, CR 2002-114, Defence R&D Canada–Suffield, 2002.